P9-EMP-239

Current Trends in Sonochemistry

Current Trends in Sonochemistry

Edited by

Gareth J. Price
School of Chemistry, University of Bath

A collection of papers presented to the Sonochemistry Symposium at the Royal Society of Chemistry Annual Congress at UMIST, Manchester, 13–16 April 1992.

Special Publication No. 116

ISBN 0-85186-365-5

A catalogue record for this book is available from the British Library

Published by The Royal Society of Chemistry,
Thomas Graham House, Science Park, Cambridge CB4 4WF

Printed by Bookcraft (Bath) Ltd.

Preface

Although applications of ultrasound in chemistry have been known for many years, the word "sonochemistry" is a relatively new addition to the chemist's dictionary. However, the number of papers featuring ultrasonically assisted chemistry is increasing annually and some processes are close to being considered viable for commercial operation on a large scale.

The first international meeting devoted to the subject was part of the Annual Congress of the Royal Society of Chemistry held at Warwick University in 1986 and the interest generated led to the formation of the Sonochemistry Subject Group of the R.S.C. in the following year. Over the past five years or so, there have been considerable developments in the subject and we felt that it was an appropriate time to hold another symposium as part of the Annual Congress, this time at U.M.I.S.T. and most of the contributors to the symposium have written papers for this volume. We were particularly pleased to have an excellent list of speakers with a strong international flavour, as reflected by the authors of the chapters in this book, and also by the fact that we attracted participants from France, Belgium, Italy, Switzerland, Portugal, United States as well as the U.K. Much of the current research work in the U.K. was also described in a poster session.

Sonochemistry is a particularly rich area of study and brings together scientists from a range of disciplines and this is reflected in the subjects in this book, ranging from fundamental physical studies of the effect of ultrasound to the synthesis of complex chemical compounds and materials. As the title suggests, the object was to summarise current activity in the subject rather than to provide a textbook. However, an introductory chapter has been included to assist the newcomer to the subject. The book is loosely organised along the lines of fundamental studies, synthetic applications and considerations for scaling up reactions or processes with commercial potential. A particular note throughout is the speculation as to where the major developments in the subject will occur in the near future.

On a personal note, I would like to thank the Congress team at the R.S.C., particularly Gina Howlett and John Gibson, for their excellent organisation at Manchester. I should also acknowledge the cooperation of all the contributors to this volume in preparing and submitting their manuscripts and also to Catherine Lyall at R.S.C. Information Services for help with the preparation of this book. I am particularly grateful for secretarial assistance to Jenny Emery of the University of Bath and to Christine Kinnear (who will be Christine Price by the time this is read. I'm also grateful to Chris for allowing this book to interfere so much with the wedding plans!).

Contents

List of Contributors

Dr G.J. Price
School of Chemistry, University of Bath, Bath, BA2 7AY, United Kingdom.

Prof. J. Reisse
Université Libre de Bruxelles, Chimie Organique, 50, Avenue F.D. Roosevelt, 1050 Bruxelles, Belgium.

Dr S. Leeman
Department of Medical Engineering and Physics, King's College School of Medicine and Dentistry, Dulwich Hospital, London SE22 8PT, United Kingdom.

Prof. J-L. Luche
Laboratoire d'Etudes Dynamiques et Structurales de la Sélectivité, Université Joseph Fourier, BP53X, 38041 Grenoble Cedex, France.

Prof. R.S. Davidson
The Chemical Laboratory, University of Kent, Canterbury, CT2 7NH, United Kingdom.

Dr C.M.R. Low
The James Black Foundation, 68, Half Moon Lane, Dulwich, London, SE24 9JE, United Kingdom.

Prof. P Boudjouk
Department of Chemistry, North Dakota State University, Fargo, ND 58105, U.S.A.

Dr J. Lindley
School of Applied Chemistry, Coventry University, Priory Street, Coventry, CV1 5FB, United Kingdom.

Dr J. Homer
Department of Applied Chemistry and Chemical Engineering, Aston University, Aston Triangle, Birmingham, B4 7ET, United Kingdom.

Prof. T.J. Mason
School of Applied Chemistry, Coventry University, Priory Street, Coventry, CV1 5FB, United Kingdom.

Dr P.D. Martin
Environmental and Process Engineering Department, AEA Industrial Technology, Harwell Laboratory, Oxfordshire, OX11 0RA, United Kingdom.

Introduction to Sonochemistry

Gareth J. Price
SCHOOL OF CHEMISTRY, UNIVERSITY OF BATH, CLAVERTON DOWN,
BATH BA2 7AY, UK

1 INTRODUCTION

The chapters in this volume, as its title suggests, are designed to describe recent developments in sonochemistry and to give up-to-date reviews of the various areas of its application. In order for the following articles to be of greater benefit to readers new to the subject, this chapter will introduce the topic in order to describe the origins of sonochemical effects, to clarify some of the terms used and to describe the types of experimental apparatus available to a chemist interested in pursuing the subject. In addition, some useful sources of information in the literature will be described. No attempt will be made to be comprehensive; rather a brief survey will be made with appropriate Literature references so as to make the subject accessible to chemists who may be interested in applying ultrasound in their own work.

In the widest sense of the word, sonochemistry refers to the study of the effects of sound on chemical reactions. However, while there are reports of chemistry being performed with sound of audible and sub-audible (*i.e.* infrasound) frequencies, by far the majority of sonochemical work has used **ultrasound**. This can most easily be defined as sound having a frequency above that of human hearing, the threshold of which is usually taken to be ~20 kHz. Ultrasound will be familiar in non-chemical situations in such guises as the navigation system of bats, sonar and medical uses. These applications, to obtain high resolution, utilise low wavelength, very high frequency (> 1-10 MHz) ultrasound of, particularly for medical applications, relatively low power. Often referred to as "diagnostic" ultrasound, this does have applications in chemistry in, for example conformational analysis, but most of the work to be described in this book will involve "power" ultrasound. For reasons of commercial equipment availability, this is usually in the range of 20 - 100 kHz, although "self-constructed" units operating up to 1 MHz have also been used for synthetic applications. These frequencies are clearly orders of magnitude lower than those associated with molecular vibrations so that any sonochemical effect must be indirect and is usually due to the action of the sound on the solvent in which a reaction is carried out.

2 ORIGIN OF SONOCHEMICAL EFFECTS

Sound, of whatever frequency, passes through an elastic medium as a longitudinal wave, i.e. a series of alternate compressions and rarefactions. This creates an acoustic pressure in the medium, P_A, which varies with time, t, according to:

$$P_A = P_{max} \, sin \, 2\pi vt \qquad (1)$$

where ν is the frequency and P_{max} the maximum pressure amplitude. Using this, we can define an acoustic intensity, I, as the energy transmitted through 1 m^2 of fluid per unit time, given by

$$I = (p_{max})^2 / 2 \rho c \tag{2}$$

where ρ is the density of the fluid in which the speed of sound is c. From a practical point of view, this definition of the acoustic intensity is not useful and we shall see later how this parameter can be measured experimentally. The intensity of the ultrasound will vary with distance, d, from its source due to attenuation caused by viscous forces and resulting in heating of the liquid. This can be represented by

$$I = I_o \, exp \, (-2 \, \alpha \, d) \tag{3}$$

where α is the absorption coefficient and depends on a range of factors such as the viscosity and thermal conductivity of the medium. At a constant temperature, the ratio (α / ν^2) must also be constant so that the attenuation is larger at higher frequencies.

Because the coupling of the sound field to the medium is never perfect, sonication causes enhanced molecular motion and, at its simplest level, this leads to very efficient mass transfer and mixing and so can enhance many chemical reactions. It also leads to the disruption of liquid-liquid phase boundaries and very efficient emulsification. However, the origin of most sonochemical effects is more complex.

Cavitation

Sample calculations using Equations (1) and (2) utilising typical values show that the acoustic pressure varies over a range of several atmospheres at kilohertz frequencies. If the negative pressures during the rarefaction phase of the waves are sufficiently large then the liquid will be "torn apart" resulting in the formation of voids or bubbles, a process known as cavitation. In practice, this occurs at pressures much less than those required to overcome the tensile strength of a liquid since there are always minute dust particles or dissolved gases present which act as nucleating sites. During its growth, dissolved gases and/or solvent vapour may diffuse into the bubble. Once formed, the bubbles can oscillate in size as the sound waves propagate but two distinct types of cavitation can be identified, although the situation is rarely unambiguously defined.

Stable Cavitation. Stable cavities exist for many acoustic cycles and may oscillate in resonance with the applied field, during which time they may grow significantly due to rectified diffusion before finally collapsing. Although it was long thought that stable cavitation had little influence on chemical systems, the large shear gradients around these bubbles give rise to many of the mechanical effects obtained on sonication.

Transient Cavitation. Transient bubbles usually exist for, at most, one acoustic cycle during which time they grow to a relatively large size, up to 150 - 200 μm, before collapsing violently and rapidly (~ 1 - 10 μs).

Although there is no doubt that the chemical effects of ultrasound are associated with cavitation, there is still discussion as to how the effects arise. Most workers apply the so-called "hot spot" theory involving the production of very high temperatures and pressures. However, Margulis *et al.* have proposed an "electrical" theory. Some sonochemical effects are similar to those produced in plasmas and this explanation is currently under review by a number of groups.

On the assumption that the bubble collapse is adiabatic, Noltingk and Neppiras, Flynn and others have shown that the maximum temperatures and pressures generated in the bubbles are given by:

$$T_{max} = T_o [P_m (\gamma - 1) / P] \quad (4)$$

and

$$P_{max} = P [P_m (\gamma - 1) / P]^{\gamma/(\gamma-1)} \quad (5)$$

where T_o is the temperature of the bulk liquid, P_m is the pressure in the bubble after collapse, P is the pressure before collapse, usually assumed to correspond to the vapour pressure of the liquid, and γ is the ratio of specific heats of the dissolved gas or vapour. Depending on the precise conditions used, solution of Equations (4) and (5) leads to maximum pressures of 1000 - 2000 bar and maximum temperatures of 4000 - 6000 K.

There is considerable experimental evidence to support this "hot spot" theory. For example, Verrall and, more recently, Suslick and co-workers have shown that the sonoluminescence induced in alkane solvents is the same as that arising from their combustion at several thousand Kelvin and also that chemical reactions such as the decomposition of metal carbonyls occurred under cavitation in the same manner as thermal processes at these temperatures.

Margulis and co-workers have shown that there are some phenomena which are not completely explained by the "hot-spot" theory and have proposed an alternative "electrical" theory. This considers the charge distribution due to dipoles in a solvent and their distribution around a cavitation bubble. Margulis has showed that during bubble formation and collapse, enormous electrical field gradients in the region of 10^{11} V m^{-1} can be generated and these are sufficiently high to cause bond breakage and chemical activity.

Whatever the precise origin of the chemical effects, we can identify three "zones" in a cavitating system: the bulk liquid, in which no *primary* sonochemical activity takes place although subsequent reaction with ultrasonically generated intermediates may occur; the centre of the bubble where the harsh conditions cause reactions in vapours and gases; and the interfacial region where there are large gradients of temperature and pressure, and also extremely high shear gradients due to shock waves and motion of solvent molecules around the collapsing bubbles. An extra factor which plays a large part in many chemical processes comes into play when cavitation bubbles form near a solid surface. Here, the collapse is non-spherical and results in jets of liquid impinging on the surface at speeds up to 100 m s^{-1}. This can result in the increase of mass transfer at the surface, the removal of oxide layers from metals, and the sweeping clean of surfaces. In the case of powdered reactants, the particles can also be set into motion at speeds of up to 100 ms^{-1} and undergo collisions which can markedly change their morphology.

Factors affecting cavitation

Since most sonochemical effects arise as a direct consequence of cavitation, the effect of changes in the reaction conditions and system parameters can usually be interpreted in terms of their effect on cavitation. A fuller discussion of these factors can be found in the references below but the main effects can be summarised here.

Ultrasound frequency As noted above, most published sonochemical work has involved frequencies of around 20 - 50 kHz since most commercially available equipment operates in this range. However, other studies performed over a wider range of frequencies have shown a definite effect, usually that, at the same intensity, cavitation, and hence sonochemical effects, is lessened at higher frequencies. This is often explained, at least qualitatively, in that the bubble has less time to grow and collapse at high frequencies so that lower final temperatures and pressures are reached and any shock waves are lessened.

Ultrasound intensity The definition of the ultrasound intensity given above is of little or no use practically and so the intensity is usually measured in terms of the energy fed into a reacting system. This can be done in terms of monitoring the electrical energy fed to the transducer (although this takes no account of its efficiency and other losses) or the movement at the tip of an ultrasound "horn". Perhaps the two most common methods are to monitor a "standard" chemical reaction, often the oxidation of iodide in carbon tetrachloride saturated potassium iodide solution (the Weissler reaction) or simply to monitor the temperature rise in a known volume of water during sonication compared to electrical heating of the same system.

In general, as might be expected, sonochemical activity rises with increasing intensity. However, there are several recent studies that show many reactions to have an optimum intensity above which the efficiency falls. An important point worth noting here is that several workers have reported, including several in this volume, that the products of a reaction can depend on the intensity and/or source of ultrasound used. Thus, it is very important that the precise conditions used for a reaction should be reported.

Dissolved gases These are important since gases can diffuse into the cavitation bubble during its growth, having two main physical effects. Equations (4) and (5) show that the conditions produced during bubble collapse depend on the ratio of specific heats of the gas. Hence, greater effects arise from systems saturated with monatomic gases than diatomics and polyatomics. However, the correlation is not perfect as other factors such as the thermal conductivity also determine the temperatures in and around the bubbles. The second effect is due to the solubility of the gas. Highly soluble gases reduce the cavitation threshold of a liquid but also lead to greater amounts entering the bubble resulting in a "cushioning" of the collapse and lessening of sonochemical activity. It should also be noted that many gases can have chemical consequences: *e.g.* oxygen is a very efficient radical scavenger and can produce apparently anomalous results.

Solvent properties Clearly, there are a large number of properties that can influence cavitational behaviour. For example, the intermolecular forces in the liquid must be overcome in order to form the bubbles. Thus, solvents with high densities, surface tensions and viscosities generally have a higher threshold for cavitation but more harsh conditions once cavitation begins. Perhaps the main factor is the solvent vapour pressure. Very volatile solvents lead to relatively high pressures in the bubble and also "cushion" the collapse so that, in most cases, high sonochemical activity is obtained in solvents with low vapour pressures or high enthalpies of vaporization.

Temperature Sonochemical systems are of great interest to the physical chemist since the usual Arrhenius behaviour of chemical reactions, their acceleration with rising temperature, is not obeyed and most reactions reported to benefit from sonication proceed more efficiently at lower temperatures. This is primarily due to the effect on the solvent properties described above.

3 EXPERIMENTAL METHODS

The generation of ultrasound relies on the conversion of electrical energy to a vibration *via* a transducer. Most transducers employ piezoelectric materials, quartz being used originally although its use has largely been superseded by more efficient ceramics such as barium titanate or lead zirconium titanate. A more recent alternative employs magnetostrictive materials such as nickel or iron cobalt alloy. A number of types of apparatus are available for introducing the ultrasound into a chemical system.

1. Direct immersion transducer. Provided that the transducer system is inert to the reaction (and *vice verca*) then it can be immersed directly in the reaction. However, the number of situations where these can be used is limited and the effects are irreproducible, particularly between laboratories. Several similar types of apparatus have been used but these are generally purpose made rather than being commercially available.

2. Ultrasound cleaning bath. A large body of sonochemical work has been performed using the type of bath used in many laboratories for routine cleaning applications and shown schematically in Figure 1.

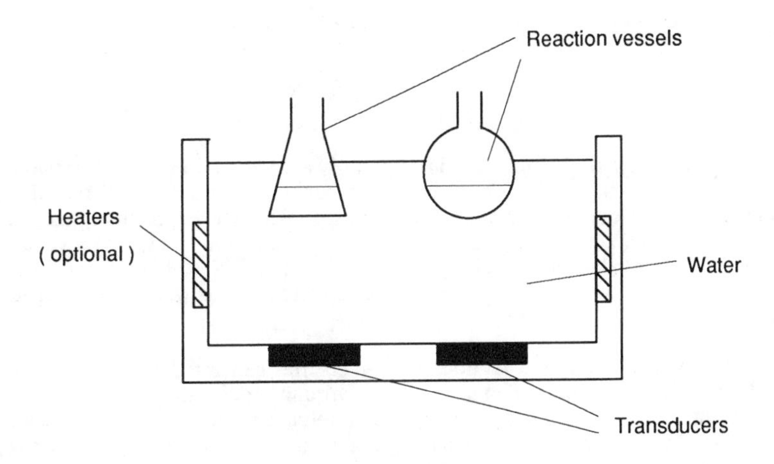

Figure 1. Ultrasonic cleaning bath.

The major advantage of this system is its economy and ready availability. However, there are several drawbacks. The ultrasonic intensity is limited by attenuation by the water and the walls of the reaction vessel. Temperature control is also difficult but can be achieved by circulating cold water through pipes immersed in the water. In general, the reproducibility of results is poor as care must be taken to ensure that the reaction vessel is in the same place in the bath for each reaction and also different baths have different frequencies and power outputs so that inter-laboratory comparisons are difficult

3. Ultrasound "horn" system. This type of system is derived from biological cell disruptors and uses a metal "horn" or "probe" attached to the transducer to amplify and introduce the ultrasound directly into a reaction or process. It is shown schematically in Figure 2. Very high intensities are available from this apparatus but care must be taken to provide adequate cooling as large temperature rises can occur. Also, the tip of the horn is subject to cavitational erosion and particles from it - usually made of a titanium alloy or stainless steel - can interfere with the chemical reaction. Fittings are available to allow reactions to be carried out under controlled atmospheres or under elevated pressures. Alternatively, a cell can be constructed to flow reactants across the end of the "horn" rather than operating in batch mode.

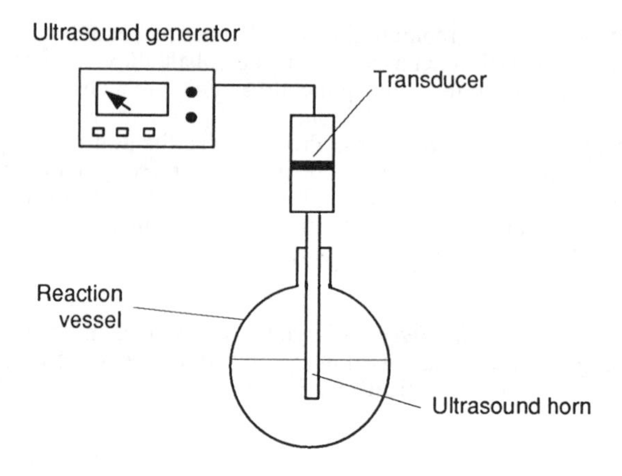

Figure 2. Ultrasonic "horn" system.

4. *"Cup horn" system.* This is essentially a combination of the last two types of equipment. A metal or glass "cup" is fitted around an inverted "horn" and filled with water into which the reaction vessel is immersed. This allows much higher intensities to be used with better temperature control than in a cleaning bath and prevents contamination by the "horn". However, the capacity of the reaction vessel is limited and the available intensity is limited for the same reasons as when using a cleaning bath.

5. *"Whistle" reactor.* Although the intensities generated by this equipment are too low to cause chemical changes, it does warrant mention since it is very efficient at producing emulsions and has been applied in polymerizations and phase transfer reactions. In this apparatus, liquids are pumped at a high rate through a narrow gap onto a thin metal blade. This sets the blade into vibration with a sufficiently high frequency to cause cavitation and hence very efficient mixing.

Over the past year or so, a number of novel systems for performing sonochemistry have become available, mainly with a view to scaling up the reactions. These are outside the scope of an introductory chapter, but several will be described later in this book.

4 LITERATURE SOURCES

Over the past five years or so, a number of books and specialist review articles covering sonochemistry have been published. In the main, these introduce the subject in much greater detail than is possible here and give extensive coverage of Literature results, particularly the applications of ultrasound in synthesis. Some of these are listed here, in no particular order, as an aid to readers new to the subject.

Books.

K.S. Suslick *Ultrasound: Its chemical, physical and biological effects* V.C.H. Publishers, New York, 1990.

S.V. Ley and C.M.R. Low *Ultrasound in Chemistry* Springer Verlag, London, 1989.

T.J. Mason *Practical Sonochemistry, A users guide to applications in chemistry and chemical engineering* Ellis Horwood, Chichester, 1991.

T.J. Mason and J.P. Lorimer *Sonochemistry, theory, applications and uses of ultrasound in chemistry* Ellis Horwood, Chichester, 1989.

Review Articles

K.S. Suslick *Science* 1990, **247** 1439.

K.S. Suslick *Scientific American* 1989, **260** 80.

J.P. Lorimer and T.J. Mason *Chemical Society Reviews* 1987, **16(1)** 239.

J. Lindley and T.J. Mason *Chemical Society Reviews* 1987, **16(2)** 453.

Also worthy of mention in this section is the annual publication *Advances in Sonochemistry* edited by T.J. Mason and published by J.A.I. Press, the third volume of which is due in the near future. Although most results are reported in the specialist synthetic, polymer or physical chemistry journals, an increasing amount of fundamental sonochemical work is also being published in the journal *Ultrasonics*.

5 CONCLUSIONS

The aim of this chapter has been to introduce sonochemistry to the beginner. I hope that all readers are now in a position to appreciate the work described in those that follow.

Quantitative Homogeneous Sonochemistry: Scope and Limitations

L. Broeckaert, T. Caulier, O. Fabre, C. Maerschalk, J. Reisse, J. Vandercammen, and D.H. Yang
UNIVERSITÉ LIBRE DE BRUXELLES, CHIMIE ORGANIQUE (CP 165), 50, AVENUE F.D. ROOSEVELT, 1050 BRUXELLES, BELGIUM

Th. Lepoint and F. Mullie
INSTITUT MEURICE, CERIA, 1, AVENUE E. GRYSON, 1070 BRUXELLES, BELGIUM

Summary : This work is essentially devoted to a systematic experimental study of the many factors which influence the rate of sonochemical reactions. A new set of sonochemical equipment working at 20 kHz is described. This equipment exhibits advantages for people interested in quantitative studies and even for people interested in the synthetic aspects of sonochemistry. Indeed, with this Chemsonic sonochemical equipment, it is possible to perform continuous and also pulsed sonication with pulses as short as 10^{-2} s.
A large part of the paper is devoted to the study of the ultrasonic field with a hydrophone coupled to a spectrum analyser. This method gives direct access to some of the primary processes which have sonochemistry as their consequences. It enables the cavitation threshold to be detected easily and also makes it possible to see how cavitation differs in water and organic solvents and how cavitation is sensitive to slight changes in the experimental conditions.
A careful study of the gas concentration during sonication was performed in order to determine if gas sequences can be safely interpreted by taking into account their physical properties and their relative solubilities at equilibrium.
Finally, a general discussion about the problems connected with absolute and effective sonochemical power is presented. It is based on comparative results obtained for the Weissler reaction on the one hand, and maleate-fumarate isomerization on the other.
A short paragraph is devoted to the "hot spot theory" versus "electrical theory" problem.

1 GENERAL INTRODUCTION

The increasing number of books, review articles, and papers devoted to sonochemistry proves that the use of ultrasound to activate chemical systems is more and more common in many laboratories. Nevertheless, it is surprising to see that the symbol))) is considered by many authors (and referees) to be a sufficient descriptor of the ultrasonic field. It is difficult today to imagine a photochemist saying that the experiment was performed in the presence of light as the only information concerning his experimental conditions !

Information about the frequency, the intensity and the type of reactor is, of course, as important in sonochemistry as it is in photochemistry. Moreover, and for reasons discussed later in this paper, many other parameters that do not play any role in photochemistry must be specified in order to define correctly the conditions of a typical sonochemical experiment. It is a serious error to underestimate the absolute necessity to define the nature of all the parameters which must be specified to describe a sonochemical experiment correctly (1,2,3,4). In the absence of such detail a sonochemical experiment described by one author will not be reproducible by another and will remain so. For us "reproducible" means "the same yield after the same reaction time" and therefore the same reaction kinetics.
To attain this goal, i.e. the definition of the parameters which have an influence on typical sonochemical kinetics, we of course need sonochemical equipment which generates well defined sonic or ultrasonic waves. In the majority of cases this condition is not fulfilled. Much commercial sonochemical equipment has been designed for other purposes (cleaning baths, cell disruptors and so on). Of course, it can be used in sonochemistry but it will rarely make possible an effective checking or even an effective knowledge of ultrasonic wave characteristics. In the best of cases, this kind of reactor gives reproducible results if the experiments are performed in the same laboratory, preferably by the same experimenter. But as soon as another cleaning bath or cell disruptor is used in another laboratory by another experimenter, different results are unavoidable and the differences may be far from negligible.

Organic chemists working with heterogeneous liquid-solid systems know how difficult it is to obtain reproducible results even in the absence of ultrasound. This is the reason why for the past few years we have centered our efforts on the Weissler reaction (4) and maleate-fumarate isomerization (2), two reactions taking place in homogeneous phase. It is interesting to compare these two reactions : the first takes place in a medium in which water is the major component while the second takes place with CCl_4 as solvent. Indeed, it is known (5,6,7) that cavitation phenomena are very different in water on the one hand, and in organic solvents on the other. Last but not least, the Weissler reaction (4) has been studied for many years and is frequently considered to be a standard.

2 A FEW WORDS ABOUT THE "CHEMSONIC" SONOCHEMICAL EQUIPMENT DEVELOPED AT BRUSSELS UNIVERSITY (ULB)

We decided to develop our own equipment in order to check as many instrumental parameters as possible. We decided to call it Chemsonic 001, Chemsonic 002. These reactors are immersion horn systems. Sandwich transducers and horns are traditional (8) in macrosonic applications with the cyclindrical horns being of aluminium or titanium. The booster effect is obtained by including a conical element in the horn shape. The resonance frequencies of our transducer-horn systems are all around 20 kHz. For each system, the frequency is known to within 1 Hz.
Two different kinds of generators have been developed which differ

according to their frequency generator systems. Chemsonic 001 is equipped with an analogic frequency generator while Chemsonic 002 has a digital frequency synthetizer. Special emphasis has been put on the generators in order to obtain a well-shaped signal which gives only one peak after Fourier analysis (fig. 1a).
A pure sinusoïdal signal gives an easy access to the electrical power delivered to the transducer. At moderate intensity, a pure sinusoïdal electrical signal permits to minimize transducer overheating and leads to a long term frequency stability in cavitation conditions. This stability itself is a prerequisite to pulsed sonochemistry, i.e. sonochemistry using short pulses but a high repetitive rate of ultrasound at a known and constant frequency.
A pulse generator adaptable to Chemsonic 001 and Chemsonic 002 has been constructed. It enables the pulse length and the interpulse waiting time to be defined independently between 10^{-2} and 99 s.

a) pure sinusoïdal signal (harmonic distortion < 0.1 %)

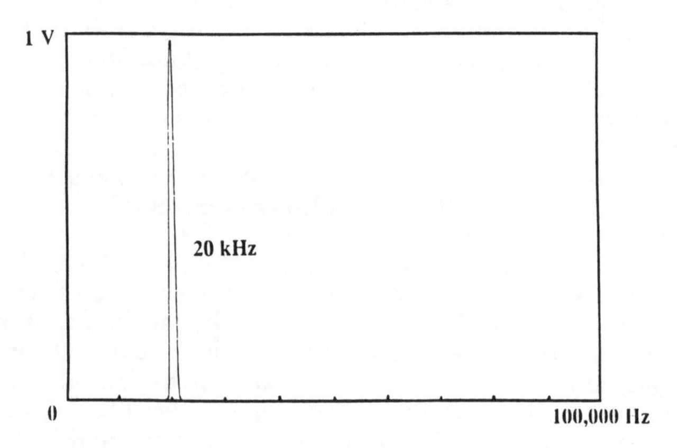

b) impure signal typical of many commercial instruments

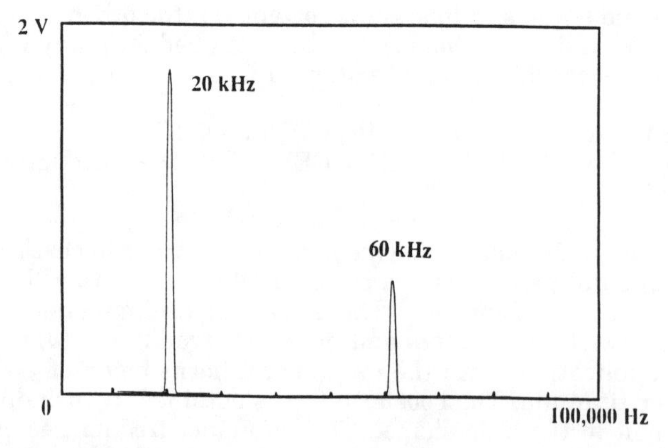

Fig. 1 : Fast Fourier Transform of the electrical signal

3 STUDY OF THE CAVITATION PHENOMENON BY AN HYDROPHONE COUPLED TO A SPECTRUM ANALYSER

Cavitation plays a central role in sonochemistry (9-12). This phenomenon, and more generally the effects of pressure waves on the liquids can be studied by using a hydrophone coupled to a spectrum analyser.
We therefore performed a systematic spectral analysis at various ultrasonic intensities of the sonic field inside and outside ultrasonic cells filled with water, cyclohexane and CCl_4. Some of these spectra are given in figures 2 and 3.

Figures 2a and 2b clearly show the presence of harmonics even at low ultrasonic intensity but the non-linear response of the system appears to be very dramatic at high intensity, above the cavitation threshold. Subharmonics ($\nu/2$, 3 $\nu/2$, 5 $\nu/2$, ..) can be observed. Of course such behaviour had already been observed, and the presence of subharmonics is sometimes considered to be direct proof that transient cavitation takes place (12,13,14).

a) low ultrasonic intensity (below the cavitation threshold)

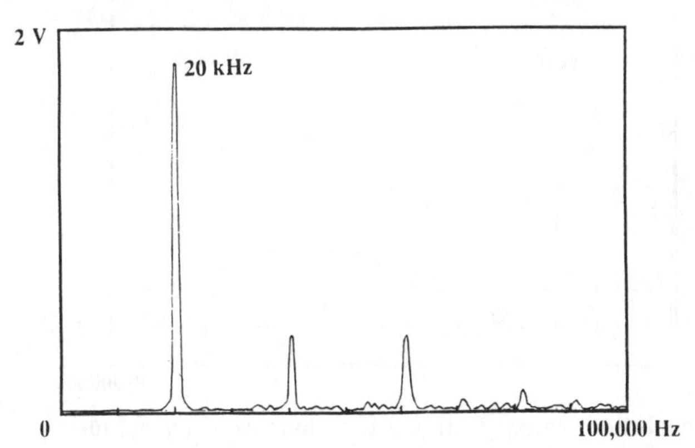

b) high ultrasonic intensity (above the cavitation threshold)

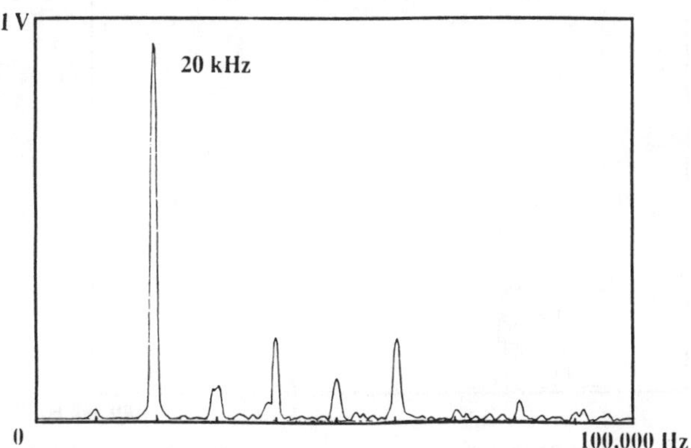

Fig. 2 : Fast Fourier Transform of the hydrophone signal(hydrophone immersed in a cell filled with water - 1V = 0.33 bar)

Figures 3a and 3b show that the same qualitative information is obtained even if the hydrophone is outside the cell. The intensities of the signals are naturally lower by many orders of magnitude when the measurement is performed outside the cell and depends on the distance between the cell and the hydrophone. Moreover the relative intensities of the signals are modified but interestingly enough, while the radial distance with respect to the cavitation zone is important, it is not the same for the angular parameters. The ultrasonic field outside the cell seems to be essentially isotropic, probably because a large number of reflections occur in the cell. It would certainly be very different with progressive waves. The possibility of obtaining the same qualitative information with the hydrophone inside or outside the cell is interesting because the hydrophone cannot be immersed in an organic solvent due to the solubility of its outer envelope in such solvents. Therefore, the study of the cavitation phenomenon in CCl_4 and in cyclohexane was performed by positioning the hydrophone externally.

a) low ultrasonic intensity (below the cavitation threshold)

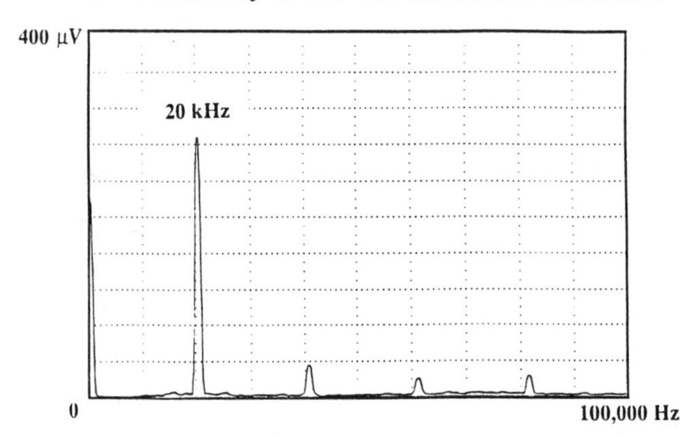

b) high ultrasonic intensity (above the cavitation threshold)

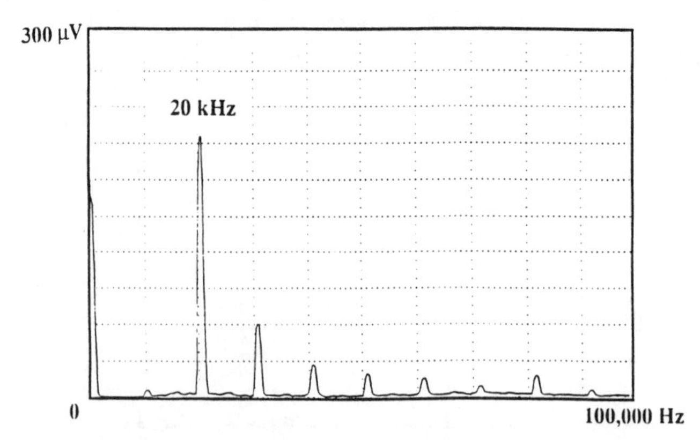

Fig. 3 : Fast Fourier Transform of the hydrophone signal (the hydrophone is outside the cell filled with water)

As is shown in figure 4a, the spectrum in cyclohexane at low intensity is qualitatively similar to what is observed in water (figure 3a). On the other hand, the comparison between cyclohexane (fig. 4b) and water (fig. 3b) exhibits large differences at high intensity (just above the cavitation threshold).
An intense noise, especially in the low frequency region, is observed in figure 4b proving that the cavitation phenomenon is qualitatively different in water and in cyclohexane. The behaviour of cyclohexane is similar to the behaviour of CCl_4. With water, it is possible to obtain spectra qualitatively similar to 4b, but much higher intensities are required in order to stay far above the threshold.

a) low ultrasonic intensity (below the cavitation threshold)

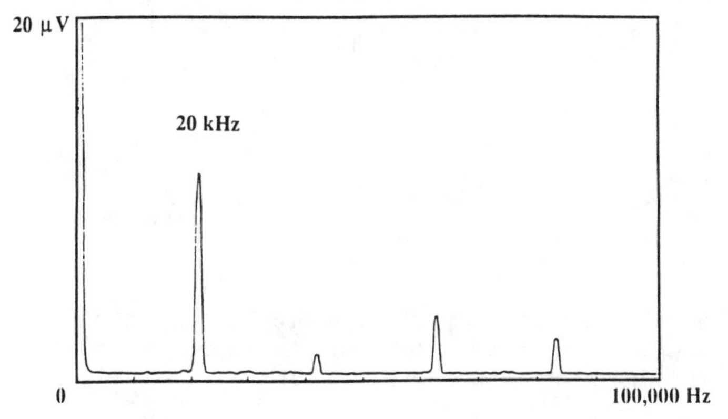

b) high ultrasonic intensity (above the cavitation threshold)

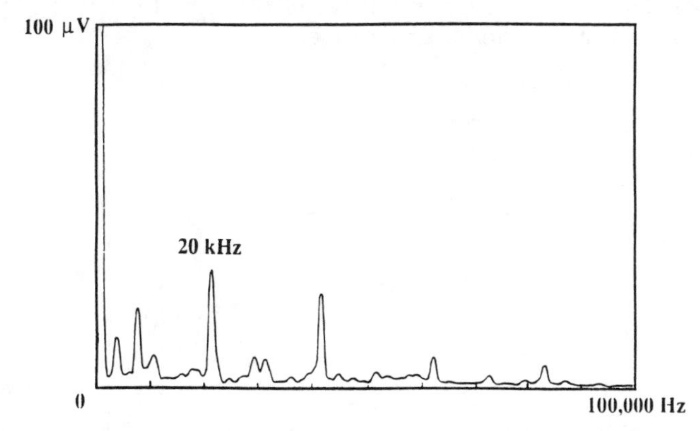

Fig. 4 : Fast Fourier Transform of the hydrophone signal (the hydrophone is outside the cell filled with cyclohexane)

Figure 5 shows a spectrum obtained with water at very low intensity (below the cavitation threshold) when the shape of the electric signal passed into the transducers is not sinusoidal and, after Fourier Transform, corresponds to figure 1b. Figure 5 must be compared to the figure 3b. Clearly, the response of a liquid submitted to a non-sinusoïdal signal is not identical to the response of this same liquid submitted to a pure sinusoïdal signal. At high intensity (above the threshold) the differences are more difficult to observe. Indeed, spectra corresponding to transient cavitation conditions are time dependent because transient cavitation itself is not a perfectly stationary phenomenon even if the ultrasound characteristics are constant.

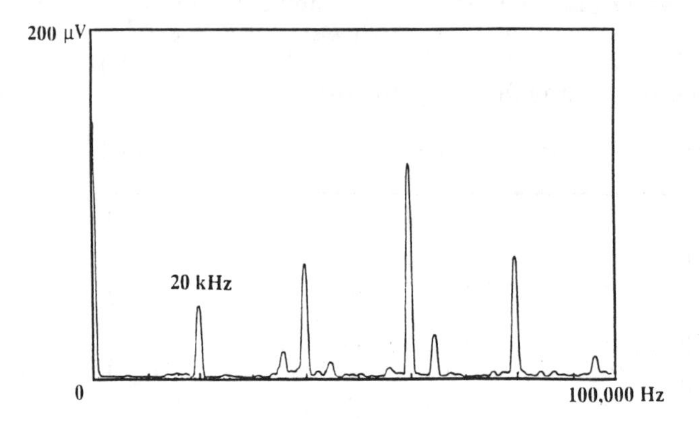

Fig.5 : Fast Fourier Transform of the hydrophone signal(same conditions as figure 3b except that the signal passed into the transducer corresponds to figure 1b)

Figure 6 shows the Fourier transform of the signal produced by a cleaning bath working in the 50 kHz range.

a) central position of the hydrophone

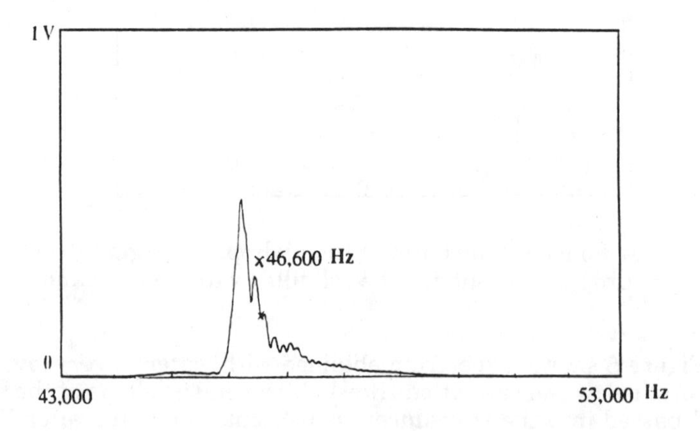

b) 1 cm decentered position of the hydrophone

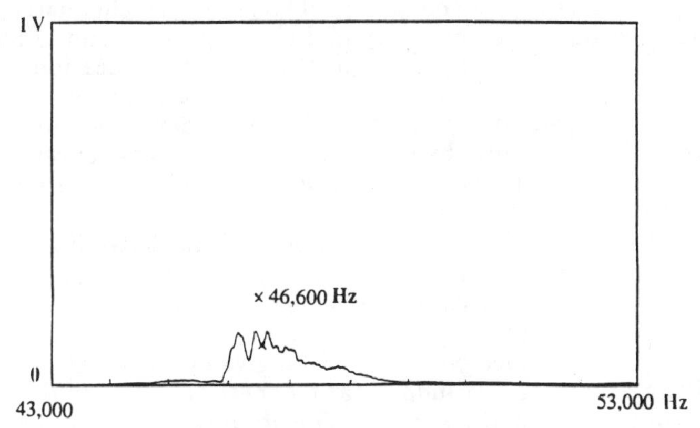

Fig. 6 : Fast Fourier Transform of the hydrophone signal (the hydrophone is inside the cleaning bath)

It is interesting to observe the difference between two spectra recorded, in the centre of the bath on the one hand, and 1 cm from this position on the other. It is certainly not easy to perform quantitative measurements under these conditions.

These few examples taken from a large body show that the physical study of the effect of ultrasound on various liquids which, by the way, are common solvents in chemistry produces interesting information for the sonochemist. We do not say that these observations can be directly correlated to yields of sonochemical reactions but they permit to compare various experimental conditions and to determine their similarities or differences.

4 QUANTITATIVE STUDY OF THE GAS CONCENTRATION IN THE SOLVENT DURING SONOCHEMICAL EXPERIMENTS. "HOT SPOT THEORY" VERSUS "ELECTRICAL THEORY"

The nature of the dissolved gas seriously affects the rate of sonochemical reactions. Based on the hot spot theory (15,16,17) it is generally accepted that monoatomic gases are more effective than polyatomic gases due to their higher Cp/Cv ratio.
We have recently shown that the situation is not so simple : CF_4 can be a better gas than argon (18), and this observation leads us to adopt a very careful attitude towards the exclusive use of the hot spot theory to explain all the chemical effects of ultrasound and also sonoluminescence (16,19,20,21). As suggested by Margulis (22,23), electrical effects could also play a significant role and even the recent sonochemical synthesis of amorphous iron from $Fe(CO)_5$ by Suslick et al (24) is not an argument in favour of the hot spot theory. It could even be considered as an argument in favour of the electrical theory. Indeed amorphous iron was already synthesized in 1983 by Lauriat (25) starting also from $Fe(CO)_5$ but putting into a plasma. As soon as the synthesis of amorphous iron from $Fe(CO)_5$ can be performed by sonochemistry in solution and by plasma chemistry in gas phase, it will be tempting to try and find other analogies between the two kinds

of chemistry and we are working in this direction. Nevertheless whatever the exact role of the dissolved gases in sonochemistry, any comparison between the efficiency of different gases initially requires the determination of the gas concentration during sonication.
Even if the solubility of two gases is similar, it is not obvious, a priori, that their concentrations will also be similar under sonication. We therefore decided to study the time evolution of the gas concentration under sonication just above the cavitation threshold at different temperatures and with CCl_4 as solvent.
Figure 7 gives a typical curve for the kinetic of the degassing of N_2. When sonication starts, the solvent is saturated with the gas. After a very short time the concentration reaches a stationary value which, in the case of N_2, corresponds to 70 % of the initial value. Same measurements performed with Ar and Xe give 60 % and 85 % respectively. It can be concluded that for these three gases the sequence of their solubilities under sonication remains the same as it is at equilibrium. This conclusion cannot be taken as general. Other similar measurements involving the common gases used in sonochemistry and the common solvents, including water, are necessary in order to prove or disprove that sequences of solubility values at equilibrium remain unchanged during various kinds of sonication procedures. The ultrasonic degassing of water has been studied very carefully by Kapustina (26,27,28).
For any quantitative study of a sonochemical reaction it is of prime importance to ensure that the gas concentration is constant during the whole experiment. If not, the type of cavitation might change during the course of the reaction (from essentially gaseous to essentially vaporous for example). It is therefore interesting to observe the presence of a plateau in figure 7.

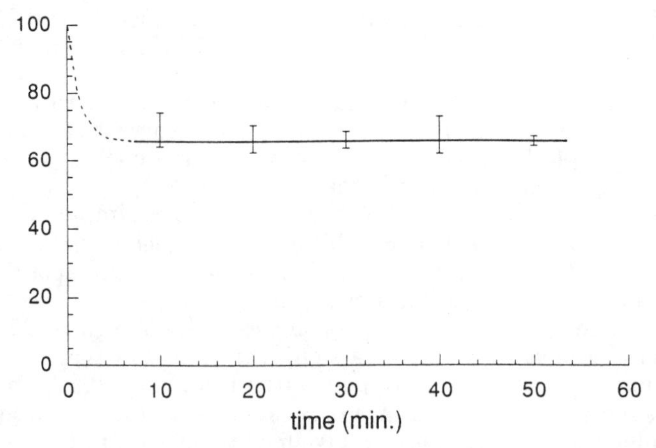

Fig.7 : N_2 relative concentration as a function of sonication time (solvent CCl_4, T=15 °C, ultrasound intensity just higher than the cavitation threshold)

A slight inverse temperature effect is observable on the diethylmaleate-diethylfumarate isomerization, with $CHBr_3$ as initiator and with CCl_4 as solvent . As soon as it is a question of the effect of temperature on the sonochemical reaction, it is of prime importance to prove that the gas concentration is the same at all temperatures. In the case of N_2 in CCl_4 it is indeed the case. The stationary concentration of N_2 remains 70 % ± 2 % of the equilibrium value in a 30°C temperature range (from -4°C to 31°C).
The negative temperature effect which we observed therefore seems to be unrelated to a gas concentration change. Nevertheless, we remain unconvinced that this conclusion is general. Any discussion if the origin of a temperature effect on a sonochemical reaction must take into account a possible gas concentration change in relation to temperature and, correlatively a change in the cavitation phenomenon.
In continuing our work on maleate-fumarate isomerization we made an intriguing observation (figure 8).

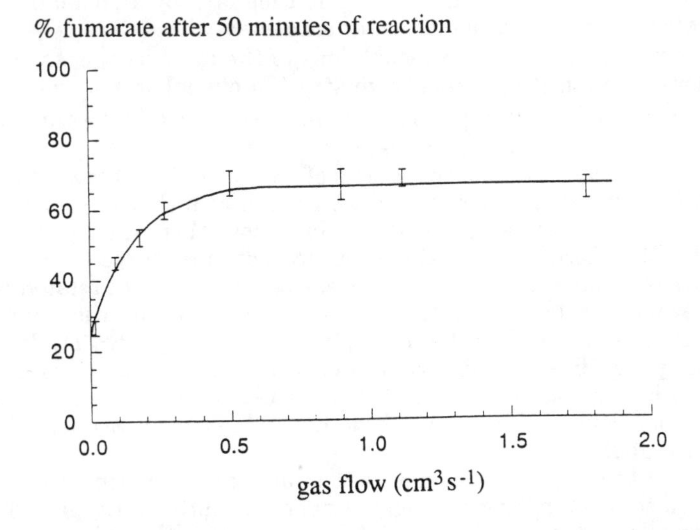

Fig.8 : Influence of the nitrogen flow above the solution on the maleate-fumarate isomerization rate (solvent CCl_4, T=15 °C)

When all the parameters are the same, the reaction rate depends heavily on the nitrogen gas flux maintained above the solution during all the sonoreaction. It must be remembered that we always start with a saturated solution. The origin of this surprising behaviour is trivial but clearly demonstrates how important the monitoring of the experimental parameters is. It is known that the presence of oxygen decreases the reaction rate (2). As soon as the nitrogen flow is insufficient (for example 0.1 cm^3s^{-1}), the concentration of oxygen, which is zero when the reaction starts, equals 8.7 % (N_2 = 91.3 %) after 5 minutes and 12.7 % (N_2 = 87.3 %) after 40.
On the other hand, when the nitrogen flux is sufficient (for example,

1.1 cm^3s^{-1}), the oxygen concentration remains equal to zero from the beginning to the end of the reaction.
We will conclude this section by saying that any study of gas or temperature effect on sonochemical reactions implies a careful quantitative study of the gas content of the solution, its time evolution during sonication, its temperature dependence. This study must be performed for each solvent, for each gas and for various irradiation intensities. A. Kapustina (27) observed, for example, that if the ultrasonic intensity is below the cavitation threshold, the degassing is less important. Any comparison between pre- and post-threshold phenomena would take this observation into account. A pre-threshold phenomenon will always correspond to cavitation conditions which are more gaseous than those taking place during a post-threshold phenomenon.

5 HOMOGENEOUS SONOCHEMISTRY WITH WATER AS SOLVENT : THE WEISSLER REACTION

The Weissler reaction, i.e. the oxidation of I^- by an oxidant generated by the sonication of water saturated with CCl_4 (homogeneous solution), is probably one of the most frequently studied reactions in quantitative sonochemistry. In our laboratory we worked with the KI and NaI/H_2O/CCl_4 system. The I_3^- concentration was measured by UV-visible spectroscopy.
We systematically studied the effects of 18-crown-6, 12-crown-4 and triglyme on reaction kinetics, and we have shown that the reaction rate is reduced in the same way by these three polyethers (figure 9).
Due to the different complexing properties of these three ethers (29), we have concluded that the rate decrease observed is not related to the complexation of the cation. It is interesting to note that the cavitation phenomenon detected by the hydrophone (see paragraph III) does not show any significant difference, if a crown ether or a glyme is added to the solution or not. At least in this particular case, the reaction rate is clearly sensitive to factors which do not influence the acoustical response of the medium.
Returning to the problems relative to quantitative sonochemistry, we observed that the hydrophone signal depends on the position of the horn in relation to the surface of the liquid of the Weissler reaction. A kinetic study generally means taking a small aliquot of the solution from time to time in order to follow the I_3^- concentration as a function of time. Therefore, as soon as a series of measurements is performed in a closed (and not in a flow) reactor, the total quantity of the solution in the reactor decreases with time. In these conditions, the position of the horn with respect to the surface changes together with the concentration of the reactants and products. It becomes necessary to demonstrate whether the change in the distance between the bottom of the horn and the surface modifies the ultrasonic intensity or not. Two different experiments were performed under exactly the same conditions : only the position of the horn in relation to the surface was changed (1.5 cm and 0.2 cm respectively). The 4 % difference between the I_3^- concentration after 30 minutes observed was slightly higher than the 2 % reproducibility of our experimental values. As has been

pointed out many times already, quantitative sonochemistry requires a great attention to many factors.

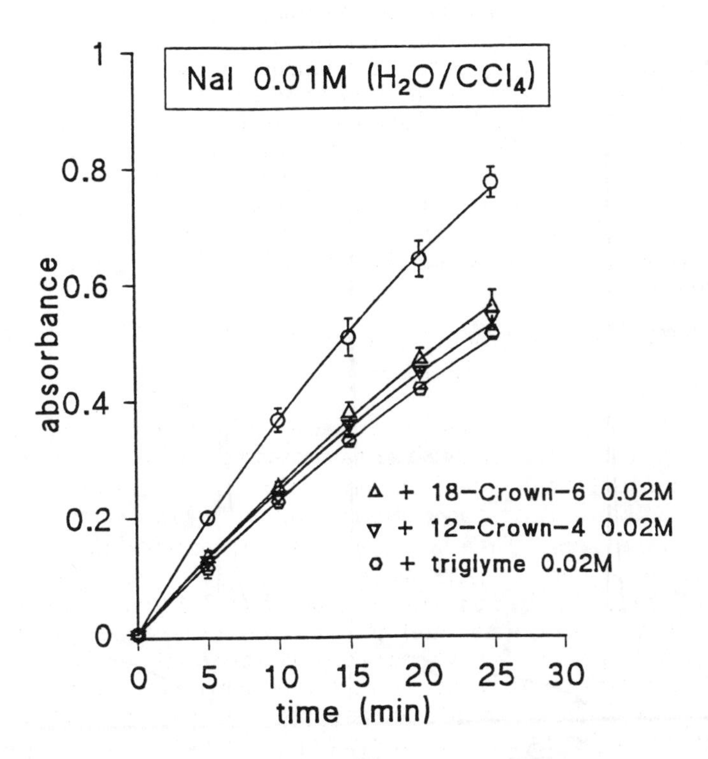

Fig.9 : Study of the Weissler's reaction in presence of crown-ethers and glyme (0.01M NaI in H2O/CCl4 mixture, 20 kHz, ultrasound intensity just above the threshold

6 A DREAM : THE ABSOLUTE INTENSITY MEASUREMENT OF THE ULTRASONIC FIELD

Everybody involved in sonochemistry would like to know the true ultrasonic intensity or power which can be used to perform reactions or to observe sonoluminescence. It is helpful to draw up a list of what can be quantitatively measured and also a list of what cannot.
With a Chemsonic immersion horn system designed to perform quantitative measurements (see paragraph II) we can select the frequency which is passed into the transducer. We know exactly what the eigen frequencies of the transducer (including the horn) are. This later piece of data can be measured by using a special device built in our laboratory. It enables a variable frequency to be sent to the transducer (including the horn system) and the electrical intensity absorbed by the transducer to be measured, at each frequency.
A plot of the absorbed intensity as a function of the frequency gives the "spectrum" of the working of the transducer-horn system. It is measured at very low intensity when exposed to air (see figure 10).

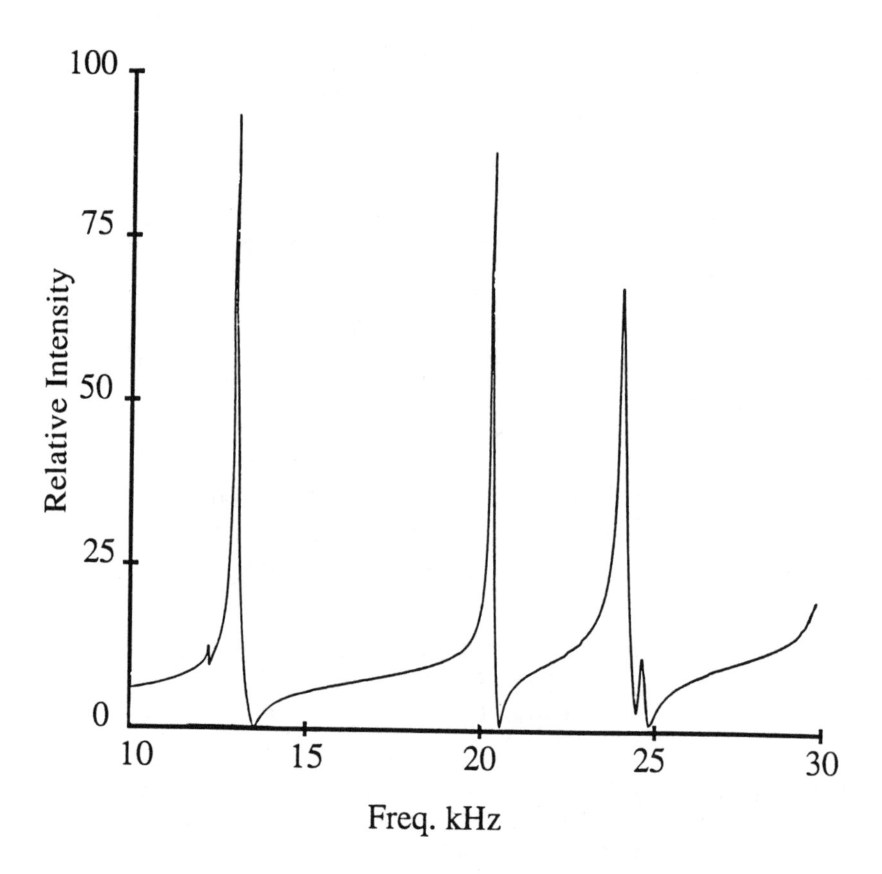

Fig.10 : " Spectrum" of the transducer horn system (example of a home-made titanium horn system)

The same measurement can be performed with the horn immersed in a solvent. At very low intensity the presence of a solvent does not significantly modify either the relative intensities of the various peaks, or their frequencies.

If the generator is tuned on the 20 kHz peak the transducer-horn systems deliver only this frequency. The small adjustments that are required after any modification in the experimental parameters can easily be carried out manually.

This special device just described is of great interest to compare various transducer-horn systems. To come back to the intensity problem, it is of course necessary to measure accurately and under all conditions the current intensity and the current voltage absorbed by the transducer.

A digital reading of these values is very helpful in determining the electrical power absorbed by the transducer
As pointed out, the accurate determination of this electrical power requires a careful control of the shape of the electrical signal. For Chemsonic 001 and 002, the maximum electrical power which can be absorbed is around 150 W.
The conversion factor of the transducer cannot be either calculated or measured accurately. Due to the absence of any significant overheating effect we are confident that the conversion factor is maximum.
The coupling between the transducer-horn system and the reactor is unknown It is certain that the reactor walls and surface of the liquid act as effective reflectors due to the very different acoustic resistance of the media in contact. The sonic and ultrasonic waves detected outside the reactor are of very low intensity (see paragraph 3). We can therefore assume that loss by radiation is small, but this assumption alone does not allow an estimate of the ultrasonic power available to perform sonochemistry (i.e. the effective power).
Many authors suggest measuring the heat effect of ultrasound to obtain an estimation of this effective power (1,30) and we did the same (2). The measurement of this heat effect can be effected by transforming the reaction cell into an adiabatic calorimeter. Nevertheless, the power so obtained is not necessarily an artifact-free measurement of the effective power. Indeed we observed that when the bottom of the horn is eroded, a very high heat production takes place but the maleate-fumarate isomerization rate decreases dramatically. In other words, heat effects are associated with cavitation but also with other mechanical effects of ultrasound, cf. those associated with streaming and shearing which are not effective (or at least less effective) for isomerization at 20 kHz.
As soon as the heat effect ceases to be a fully satisfactory measure of sonochemically effective power, it becomes doubtful whether other indirect data like the erosion effects of ultrasound can be considered as a better descriptor. Indeed, the mechanical and chemical effects of ultrasound do not depend, on the cavitation phenomenon in the same way.
We are forced to consider that probably the best (but far from perfect) solution to the problem is the choice of a reference reaction for which the rate can easily be measured. This means, for example, that all the reaction yields after a well defined sonication time could be expressed as relative values with respect to the yield of the reference reaction after the same reaction time.
The situation is far from perfect because it is not possible to find a standard reaction which can be performed in all solvents. We compared the rate of the Weissler reaction at two frequencies and various intensities (31) (see figure 11), using in this particular case the Undatim Ultrasonics 20 kHz generator equipped with two different horns (a titanium horn and a steel horn).

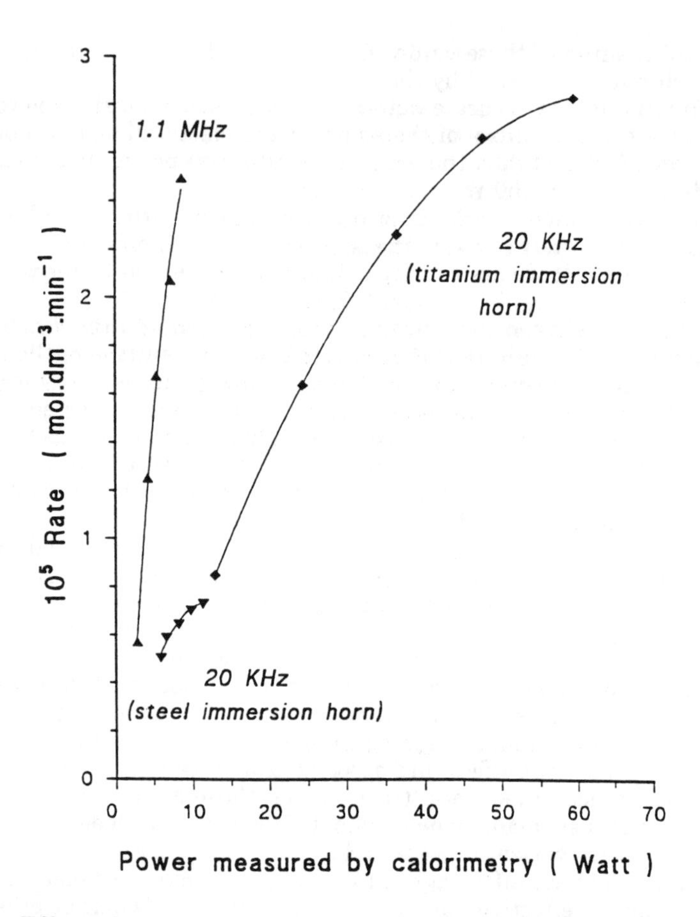

Fig.11 : Effect of freqency and ultrasound intensity on the Weissler reaction

It is clear that for the same power measured by calorimetry the reaction is much faster at high than at low frequency (32,33). We conclude that calorimetry does not give a good estimation of sonochemical effective power as long as we compare two frequencies. For a same frequency, the correlation between the rate constant and the heat effect is acceptable and can thus be used as soon as the horn is not eroded. Nevertheless, even if we take the rate of the Weissler reaction at the two frequencies as references and compare the rate of a particular reaction (say the isomerization of maleate in CCl_4) for the same values of the power supply or for the same values of the heat effects, we already know that the sonochemical effective powers will not be the same due to the solvent difference (water in one case, CCl_4 in the other case). Indeed we observed that with our apparatus and in the accessible power supply ranges, the isomerization reaction does not take place at high frequency while it is easily performed at low frequency. This observation clearly illustrates that it is impossible to define a reference reaction valid in all circumstances and especially for various frequencies.

7. GENERAL CONCLUSION

This paper is entirely devoted to a systematic study of some of the parameters which affect the rates and yields of sonochemical reactions. Many of these parameters can be checked with a satisfactory degree of accuracy and nothing precludes their systematic laboratory checking provided that the equipment fulfils the necessary requirements. Nevertheless, the absolute and even relative measurement of sonochemically effective power (or energy) remains an open problem and limits quantitative studies or at least comparisons between quantitative studies performed with different equipment. It is doubtful that this problem will be solved easily and it remains the most important challenge in quantitative sonochemistry. It is via the coupling of different chemical and physical techniques including direct measurements by hydrophone that this problem will probably be solved.

ACKNOWLEDGEMENT

This work was supported by a grant of the "Fonds National de la Recherche Scientifique" and by "La Loterie Nationale".

EXPERIMENTAL PART

Characteristics of home-made equipment

Chemsonic 001
Analogic Frequency Generator 18-22 kHz
Stability ± 10^{-4}
Total harmonic distorsion 0.5 %

Chemsonic 002
Digital Frequency Synthesizer, 5 Hz to 200 kHz by steps of 1 Hz or 10 Hz
Stability ± 10^{-5}
Total harmonic distorsion, at 20 kHz - after filtering 0.1 %

Power amplifier (identical for Chemsonic 001 and 002)
Frequency response 5-100 kHz
RMS power 150 W with a total harmonic distorsion of 0.1 % at 20 kHz

Pulser : quartz controlled from 00.1 to 99 sec. range

Transducer
2 stacked Philips piezoceramics of polycristalline $PbTiZrO_3$
Aluminium or titanium horn

Characteristics of commercial equipment

Undatim Ultrasonics equiment
Frequency 20 kHz. RMS power 300 W

Ultrasonic cleaning bath Bransonic 220
Frequency range : around 50 kHz.

Dynamic Signal Analyser Hewlett Packard model 3516A
Single channel Fast Fourier Transform (FTT) covering the 0-100 kHz range
Acquisition time : 4.10^{-6}s
The spectra shown in the figures correspond to averages over 10 scans.

Hydrophone
Brüel and Kjaer miniature hydrophone, type 8103 (diameter : 10 mm, length : 30 mm). Frequency range in water 0.1 to 100 hKz. Sensitivity 26.9 μV/Pa.

Experimental conditions

Isomerization of diethyl maleate to diethyl fumarate
Solvent CCl_4, initial maleate concentration 0,21 M, initial bromoform 0,38 M concentration, inert gas N_2.
Quantiative analysis by gas chromatography.

Weissler reaction
Solvent : water saturated with CCl_4 (aerated solution).
Initial I^- concentration 0,1 M : Quantitative analysis by UV-visible spectroscopy.

Gases analysis.
All the analysis were performed by GC (IGC 120) on Carbowax 20M (10% on Chromosorb P) and on molecular sieves 5A, 60-80 mesh for the N_2-O_2 mixture.

REFERENCES

1. T.J. Mason, J.P. Lorimer, D.M. Bates, Ultrasonics, 1992, 30, 40
2. J. Reisse, D.H. Yang, M. Maeck, J. Vandercammen and E. Vander Donckt, Ultrasonics, accepted for publication, 1992
3. B. Pugin, Ultrasonics,1987, 25, 49-55
4. A. Weissler, H.W. Cooper and S. Snyder, J. Am. Chem. Soc., 1950, 72, 1769
5. P.I. Golubnichii, V.D. Goncharov and K.V. Protopopov, Sov. Phys. Acoust., 1970, 15(4), 464-471
6. P.I. Golubnichii, V.D. Goncharov and K.V. Protopopov, 1971, Sov. Phys. Acoust.,1971, 16(3), 323-326
7. A. Grivnin Yu, S.P. Zubrilov and V.A. Larin, Russ. J. Phys. Chem., 1980, 54(1), 30
8. J. Gallego-Juarez in "Power Transducers for Sonics and Ultrasonics", B.F. Harmonic, O.B. Wilson and J.N. Decarpingny (Editors), Springer Verlag, Berlin, 1991, p. 35
9. I.E. Elpiner, Ultrasound, "Physical, Chemical and Biological Effects", Consultants Bureau, New York, 1964

10. M.G. Sirotyuk in "High-Intensity Ultrasonic Fields", L.D. Rozenberg (Editor), Plenum Press, New York, London, 1971, p.263
11. L.D. Rozenberg in "High-Intensity Ultrasonic Fields", L.D. Rozenberg (Editor), Plenum Press, New York, London, 1971, p.347
12. A.A. Atchley and L.A. Crum in "Ultrasound, its Chemical, Physical and Biological Effects", K.S. Suslick (Editor), VCH, New York, 1988, p.1
13. F.G. Sommer, D. Pounds, Med. Phys., 1982, 9, 1
14. J.A. Rooney in "Ultrasound, its Chemical, Physical and Biological Effects", K.S. Suslick (Editor), VCH, New York, 1988, p.65
15. V. Griffing, J. Chem. Phys., 1950, 18, 997
16. V. Griffing, J. Chem. Phys., 1952, 20, 939
17. M.E. Fitzgerald, V. Griffing, J. Sullivan, J. Chem. Phys., 1956, 25, 926
18. Th. Lepoint, F. Mullie, N. Voglet, D.H. Yang, J. Vandercammen and J. Reisse, Tet. Letters , 1992, 33, 1055-1056
19. K.S. Suslick, Science, 1980, 247, 1439
20. R.E. Verrall and C.M. Sehgal, in "Ultrasound, its Chemical, Physical and Biological Effects", K.S. Suslick (Editor), VCH, New York, 1988, p.277
21. F.B. Flint and K.S. Suslick, J. Am. Chem. Soc., 1989, 111, 6987
22. M.A. Margulis, Russian J. Phys. Chem., 1985, 59, 882
23. M.A. Margulis, Advances in Sonochemistry, 1990, 1, 39 Ed. T.J. Mason, JAI Press LTD.
24. K.S. Suslick, S.B. Choe, A.A. Cichowlas and M.W. Grinstaff, Nature, 1991, 353, 414
25. J.P. Lauriat, J. of Non-Crystalline Solids, 1983, 55, 77
26. O.A. Kapustina, Soviet Physics (Acoustics), 1964, 9, 346
27. O.A. Kapustina, Soviet Physics (Acoustics), 1965, 10, 376
28. O.A. Kapustina, Soviet Physics (Acoustics), 1970, 15, 328
29. C.J. Pedersen in "Synthetic Multidentate Macrocyclic Compounds" Ed. R.M. Izatt and J.J. Christensen, Academic Press, 1978, p.1.
30. M.A. Margulis, A.N. Mal'tsev, Russ. J. Phys. Chemistry, 1969, 49, 592
31. J. Reisse, oral presentation at the 2d Symposium of the European Sonochemical Society (Gargnano, Italy, September 91).
32. M.A. Margulis and Y.T. Didenko, Russ. J.Phys. Chem., 1984, 58, 848
33. C. Petrier, A. Jeunet, J.L. Luche and G. Reverdy, J. Am. Chem. Soc., 1992, 114, 3148

Cavitation Phenomena

S. Leeman and P.W. Vaughan
DEPARTMENT OF MEDICAL ENGINEERING AND PHYSICS, KING'S COLLEGE SCHOOL OF MEDICINE AND DENTISTRY, DULWICH HOSPITAL, LONDON SE22 8PT, UK

1 INTRODUCTION

If a liquid sample is exposed to an ultrasound field, it is widely observed that, under appropriate conditions, a variety of chemical, physical, and biological phenomena may be activated. Some of the effects occur at definite thresholds in the intensity of the applied ultrasound field, and it is those which are conventionally associated with (acoustic) cavitation. It is commonly sought to utilise such cavitational effects as a means for monitoring the presence and strength of the cavitation itself, but problems of consistency arise, as the observed thresholds are not necessarily all coincident. Clearly, before a cavitational monitoring scheme can be implemented, it is essential to have a clear definition for the phenomenon itself. Rather surprisingly, the literature abounds with a number of definitions, not all specific, and some mutually contradictory. A primary aim must therefore be to establish an unambiguous definition for cavitation.

It is generally agreed that 'bubbles' in the sonicated liquid play a fundamental role in the cavitational process. In particular, the so-called 'Rayleigh cavity' - or, more accurately, 'void' - has played a crucial part in the development of the notion of transient cavitation. It is interesting to note, however, that Rayleigh, in his pioneering paper,[1] makes no claim as to the originality of the model, which he ascribes to Besant. The concept was introduced specifically in the context of hydrodynamic cavitation, and it is by no means clear that it is relevant to acoustic cavitation. Moreover, Rayleigh himself considered that a gas-filled bubble, rather than Besant's void, was a more reasonable physical probability. In that case, he showed that a 'catastrophic' collapse was not likely: rather, the bubble would rebound from its minimum volume configuration, as a result of the cushion of gas trapped within. However, both direct observation of acoustic cavitation phenomena,[2] as well as theoretical models,[3,4] have strongly emphasised the primary role that (gas- or vapour- filled) bubbles play in the cavitational process, and it seems reasonable to assume that the phenomenon would not occur in their absence.

We have been led to define acoustic cavitation as the nonlinear pulsation of gas- or vapour- filled bubbles in an acoustic field. The emphasis on the nonlinearity of the pulsation is essential in order for the definition to be in accord with the conventional notion that cavitation has a definite intensity threshold, and is consistent with the predictions of bubble-theoretic models. The implication is that, contrary to some current (and, indeed, unacceptable) definitions, any bubble activity is not to be regarded as cavitational, and it is irrelevant whether or not the bubbles are actually generated by the field. Moreover, it is immediately apparent that an appropriate method for the detection of cavitation could be based on the monitoring of acoustic emissions characteristically associated with nonlinear bubble pulsations.

2 STAGES OF ACOUSTIC CAVITATION

Consider the behaviour of a bubble in a liquid that is exposed to an ultrasound field whose intensity is monotonically increased. At very low applied field powers, the bubble will pulsate linearly, in synchrony with the local acoustic pressure. Such activity has no definite threshold, in general, and is hardly expected to give rise to the manifold of effects conventionally associated with cavitation; by the definition posited above, it is, in fact, a non-cavitational process. As the applied power is increased, the amplitude of the bubble pulsation will increase, and will be driven into a nonlinear mode (i.e., the pulsation will assume frequency components that are not present in the driving field). In accordance with the general theory of such processes, nonlinearity will be induced at a well-defined intensity threshold - which, in this case, also heralds the onset of 'cavitation'.

It is instructive to focus attention on the bubble wall velocity. Close to the cavitation threshold, the maximum wall velocity at any location in the bubble, U_{MAX}, will not exceed the sound velocity, C_{GAS}, in the vapour or gas in the bubble's interior. We have called this the "subsonic" stage of cavitation, which is characterised by the presence of fractional harmonics in the acoustic emission from the bubble. At higher power levels, the bubble wall velocity may exceed C_{GAS}, and the potential for micro-shock formation inside the bubble is realised. We have suggested that this ("gas phase") cavitational stage marks the onset of sonoluminescence, as well as some gas-phase sonochemical reactions. At even higher applied powers, U_{MAX} will exceed the velocity of sound in the ambient liquid, with shock waves propagating in the liquid. This is the "liquid phase" stage of cavitation, and is probably best characterised by associated (liquid phase) sono-chemical reactions.

It will be noted that the above view of cavitation is in accord with a relatively low threshold for sub-harmonic emission from gassy liquids (as observed over a wide frequency range), and insists that this lies lower than the threshold for sonoluminescence (as observed in a number of experimental studies). It also implies that a possible hierarchy of sonochemical reaction 'thresholds' may be associated with increasing levels of applied ultrasound intensity. It de-emphasises concepts such as 'stable' and 'transient' cavitation, which are poorly and ambiguously defined, and whose conventional association with 'gentle' and 'violent' cavitational processes has no clear experimental justification.

The three-stage view of cavitation is remarkably successful in explaining the relative strengths of subharmonic and sonoluminescent emissions from water saturated with a variety of gases,[5] and sonicated with 1MHz or 1.5MHz continuous wave fields. More recently, we have conducted a number of experiments in a modified commercial (Branson) cleaning bath, and have been able to confirm the general validity of these conclusions also for a pulsed, 50kHz field. The experimental setup is schematically shown in Fig 1.

The water sample was carefully degassed, then saturated with a selected gas at a controlled temperature, in such a way that contamination with air was avoided. The spectrum of the acoustic emission over the range 0-200kHz, sonoluminescence output, and the erosional effects on a thin aluminium foil (for fixed time of exposure to the field) were all simultaneously monitored for a number of drive powers. It is informative to consider the results obtained with the gases He, Ar, and CO_2, which can be summarised as follows:

Gas	Fractional harmonics	Sono-luminescence	Erosion
He	*****	*	*****
Ar	***	*****	***
CO_2	*	*	*

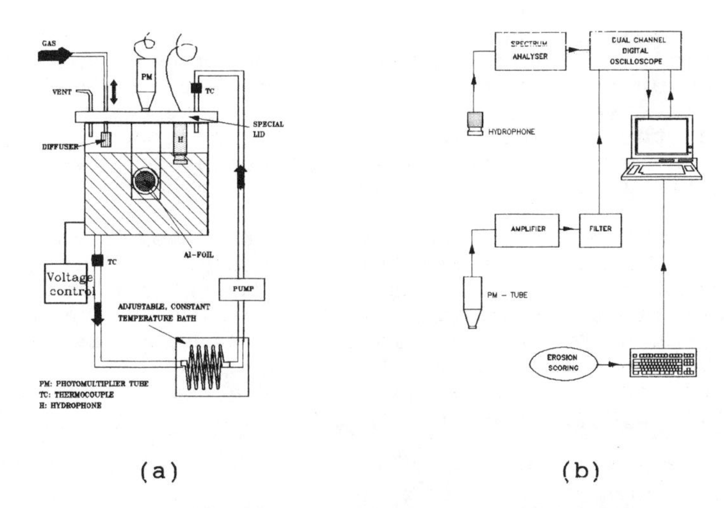

Figure 1 Schematic diagram of experimental setup showing (a) the modified Branson cleaning bath, and (b) the data acquisition pathways.

Only qualitative results need be reported here: the more strongly a gas acts to produce a given effect, the more stars are indicated. He and Ar appear to have quite opposite effects as regards their fractional harmonic emission, sonoluminecence, and erosional capabilities, while CO_2 is very weak indeed in all these capacities. Acoustic emission and sonoluminescene were quantitatively monitored, but the erosion was based on a scoring system which emphasised the total area of noticeable holes punched in the foil. In all experiments, the acoustic emission threshold, for any given sample, was always observed to be lower than those for sonoluminescence and erosion - in accord with the definition and three-stage model for cavitational processes outlined above. Clearly, the mechanisms which give rise to sonoluminescence do not parallel those that lead to copious acoustic emission, the latter being more closely coupled to erosional processes.

In order to understand the above results in terms of the three-stage model, the properties of the dissolved gases that influence cavitational effects[5] obviously need to be taken into account. The acoustic (fractional harmonics) emission at a given drive intensity depends on the strength of the nonlinear pulsations, i.e. the ease with which large amplitude pulsations can be excited. This will occur the more (effectively) isothermal - or less adiabatic - the bubble dynamics are. Thus the higher the thermal conductivity of the gas, the more actively the bubble will emit fractional harmonics. The higher its solubility, the more gas is available to 'leak' from solution into an expanding bubble, and the more effectively the pulsation can be damped during contraction: thus the acoustic emission is reduced. Sonoluminescence output depends on the temperature of the bubble contents ('hot spot' theory[4]), and is therefore reduced the more isothermal the motion. Thus, high thermal conductivity of the gas inhibits sonoluminescence. Moreover, 'solubility damping' of the motion implies that high bubble wall velocities cannot be easily attained, so that high gas solubility will effectively drive the sonoluminescence (intensity) threshold up. Clearly, all other conditions being equal, the higher C_{GAS}, the higher the sonoluminescence threshold (and the weaker the light output at a given drive intensity, above that threshold). The ratio of the specific heats of the gas, γ, also influences the bubble dynamics: the lower γ, the more effectively isothermal the motion, and thus the lower the sonoluminesence threshold. But it can be shown that, as γ increases, the more effective the (micro-) shock wave heating of the bubble's interior becomes, and the more intense the sonoluminescence.[5,6]

For the gases considered here:
Thermal conductivity: He >> Ar, CO_2
Solubility: CO_2 >> Ar, He
C_{GAS}: He >> Ar, CO_2
γ: Ar, He > CO_2
It is readily ascertained that the relative intensities of both the fractional harmonic and sonoluminescent emissions from water saturated with these gases are very much as qualitatively expected within the context of the three-stage cavitational model. Unexpected predictions of the model that have been verified, include the observation that the sonoluminescence threshold for N_2-saturated water is lower than that for Ar, even though the latter is a stronger sonoluminescing agent[6]. Moreover, it is apparent that at very low gas concentrations, the solubility damping effect should be very much reduced: indeed, this has been found to be the case with CO_2, which will generate both substantial fractional harmonics and sonoluminescence when present in only low concentration. Note that in arriving at an understanding of the above phenomena, no resort to the concepts of stable/transient cavitation has been necessary.

3 A CAVITATION INDEX

In many applications, a measurable parameter that indicates the 'amount', or 'intensity', of the cavitation occurring is required. In some sense, this would indicate the 'cavitational efficiency' of the applied acoustic field, and should be in rough accordance with intuitive notions about the degree of cavitational activity. A number of cavitational effects may be considered as possible candidates for a cavitational index:[7] acoustic emission, sonoluminescence, chemical effects, changes in the properties of the liquid medium (such as density), and erosion rate of a thin metal foil. Since these effects are all influenced to some degree by a number of circumstances other than the cavitational intensity itself, it is appreciated that the the search for a cavitational index is beset with difficulties. The problem is highlighted by the results reported above: Ar-saturated water would indicate intense cavitational activity on the basis of sonoluminescence, but, under the same conditions, would indicate only moderate activity for an index based on acoustic emission. In He-saturated water, the reverse situation would be seen to hold!

In order to arrive at a meaningful cavitational index, it is desirable that the index be linked directly to an unambiguous definition for cavitation itself. Moreover, the consistency of the index should be maintained over a range of applied powers, temperatures, frequencies, and, at least, for a variety of dissolved gases. In this context, foil erosion is generally regarded as a reasonable indicator of cavitational activity: indeed, were it a more convenient, reproducible and quantitative measure, it would probably form the basis for a suitable index. On the other hand, it is known that erosion can be inhibited under certain circumstances, and it is difficult to see how it relates directly to a definition for cavitation.

It should be emphasised that the appearance of the foils indicates that a variety of 'erosion' processes are active. The wide range of effects seen suggests that, as with all cavitational phenomena, a rather complex situation exists. In particular, there is clear evidence that distinctly different processes are operative in degassed and gassy water. This is interpreted by us to be further experimental verification of the concepts of 'vaporous' and 'gaseous' cavitation, as first postulated by Blake.[2] The existence of these two categories is entirely consistent with the three-stage cavitational approach outlined here. Erosion of the foils is scored in a semi-quantitative way, such that it reflects primarily the area of noticeable holes: when assessing results, it should be borne in mind that reproducibility is only moderate (±30%)

Given the definition for cavitation posited above, it is only natural that a proposed cavitation index should be based on the strength of the fractional harmonics, relative to the drive intensity. In practice, we have found that an index taking into account only the first half-harmonic (at frequency $f_D/2$, with f_D the drive frequency) is adequate. The following cavitational index is therefore suggested:

$$\frac{strength\ \ of\ \ f_D/2\ \ subharmonic}{strength\ \ of\ \ f_D\ \ (fundamental)}$$

Note that higher harmonics are specifically excluded from the index, because, although they are also associated with nonlinear bubble pulsation, they may also be a consequence of nonlinear (drive) wave propagation - even in the absence of bubbles. Fractional harmonics, on the other hand, are closely bound up with the existence of bubbles, and their nonlinear pulsation.

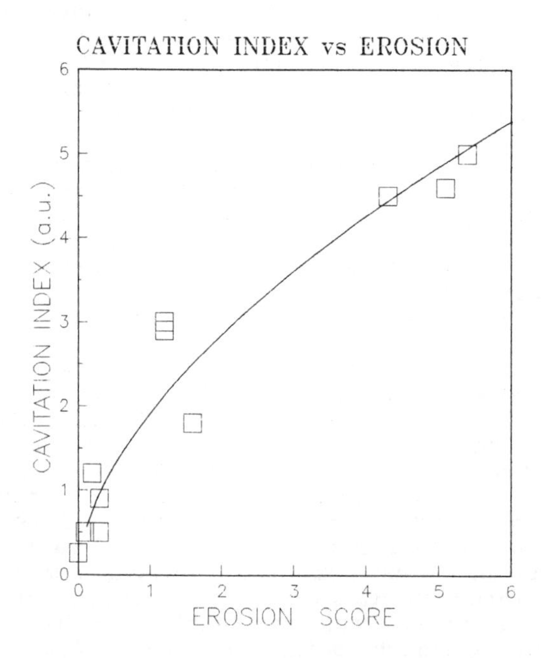

Figure 2 Correlation between erosion score and cavitation index (arb. units) for He-, Ar-, and CO_2- saturated water, at 20°C, and for a variety of drive field intensities.

Results of experiments carried out at 20°C are summarised in Fig. 2. A cavitation index based on sonoluminescence is clearly inadequate, as it does not consistently rise with increasing erosion, irrespective of the dissolved gas. On the other hand, the proposed index that is based on subharmonic emission performs well, showing a consistent correlation with erosion, irrespective of the dissolved gas: it thus appears to be less sensitive to the actual ambient conditions, and more directly related to cavitational intensity itself. A cavitation index based on acoustic emission has other strong advantages: it allows cavitation to be continuously monitored, in a remote fashion, even when occurring in possibly hostile, optically opaque, environments. Moreover, it relates directly to a fundamental definition for the cavitation process itself, and may therefore be reasonably expected to remain a consistent index even when erosion is compromised.

4 CONCLUSIONS

We have demonstrated that a clear and unambiguous definition for acoustic cavitation is possible. The definition implies a three-stage description of cavitation, which obviates the need to have recourse to poorly justified concepts such as stable, transient, and 'transient-violent' cavitation. It also enables an uncomplicated and consistent approach to fractional harmonic and sonoluminescent activities to be devised, and implies the potential exists for possibly diverse sonochemical reactions to be stimulated in two different cavitational stages. A meaningful and measurable cavitation index, based on the strength of subharmonic emissions, may be devised.

REFERENCES

1. Lord Rayleigh, Phil. Mag., 1917, 34, 94.
2. F.G. Blake, 'Tech. Memo No. 12', A.R.L., Harvard University, 1949.
3. E.A. Neppiras, Phys. Repts., 1980, 61, 108.
4. F.R. Young, 'Cavitation', McGraw-Hill, London,1989.
5. P.W. Vaughan and S. Leeman, Acustica, 1989, 69, 109.
6. P.W. Vaughan and S. Leeman, Acustica, 1986, 59, 279.
7. H.G. Flynn, In: 'Physical Acoustics', 1B, Ed. W.P. Mason, Academic Press, New York, 1964.

New Orientations in Sonochemistry

J.L. Luche
LABORATOIRE D'ETUDES DYNAMIQUES ET STRUCTURALES DE LA SÉLECTIVITÉ, UNIVERSITÉ JOSEPH FOURIER, BP 53X, 38041 GRENOBLE CEDEX, FRANCE

1 INTRODUCTION

The essential purpose of chemistry is to provide compounds to users, and to learn how to make them. Because of the necessities of economic savings, improved technology and now, environmental requirements, chemists have been engaged for a long time in looking for new methodologies able to achieve chemical transformations under optimum conditions. Before the recent development of biotechnology, physical agents have been almost exclusively employed to perform reactions; heat and pressure being followed by light and electricity. It seems now that physical agents of a third generation should come under study, not only for fundamental curiosity, but also with the aim of possible industrial applications. Among these new chemistries, sonochemistry has taken a particular place. From an empirical state of the art, a more rational knowledge has been obtained in recent years, which now permits us to predict its domains of operation with some chance of success. Synthesis is obviously one of them, but we can envisage that mechanistic studies using sonochemical methods will develop in the future. Since there may exist some links between sonochemistry and newly developing unusual chemistry, a short discussion illustrated by examples of mechanochemical reactions will be given.

2 SONOCHEMISTRY IS NOW A NEW FIELD IN CHEMISTRY

Until recently, ultrasonic irradiation was considered as a simple laboratory trick, with uses limited to that of improved stirring, or as an exotic field of chemistry with a simple phenomenological interest. Even if their work is of considerable value from a theoretical point of view, Elpiner[1] then Margulis[2] in Moscow, and Henglein[3] in Berlin studied systems which did not, at the time, stimulate applied research and general uses. Possibilities in the synthesis of several organometallics were perceived as soon as 1950 however, by Renaud.[4] Almost 30 years later, Fry[5] showed that α,α'-dibromoketones could be brought to reaction in an original manner by sonication in the presence of mercury as shown in Figure 1.

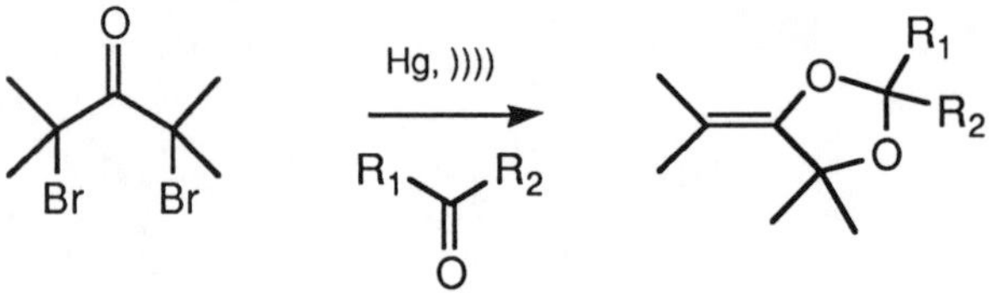

O
Br Br
Hg,))))
R_1 R_2
O
R_1
O
R_2
O

Figure 1: A mercury sonoreaction.

Unfortunately, chemists, especially in industry, do not like mercury, and sonochemistry did not develop from these initial studies. The situation changed at the beginning of the 1980's, with several papers in organometallic synthesis,[6] homogeneous organic reactions,[7] and transition metal chemistry.[8]

The development however is still more or less empirical, because the physical phenomena by which ultrasound interacts with matter are poorly understood. Since an important obstacle rises when a theoretical route is followed, it seems safer to rely on experimentation to determine the common trends amongst the reactions reported to be sensitive to ultrasound. The formulation that we have reached is the following:[9]

1:- In homogeneous solution, the interaction of ultrasound with a reactive system induces monoelectronic transfers. Transition metal complexes undergo ligand-metal bond cleavage, yielding coordinatively unsaturated species. Purely ionic reactions should not be sensitive to sonication.

2:- In heterogeneous liquid-liquid or liquid-solid systems, ionic reactions can be stimulated by mechanical effects. Rates and yields can be improved depending on the physical parameters which characterise the biphasic system: surface tension, density, mechanical properties of the solid. The nature of the products obtained under these conditions will, however, be the same as in the absence of ultrasound.

3:- In heterogeneous systems, reactions which can follow an ionic or an electron transfer pathway will be induced to react preferentially according to the second mode. The mechanical component will obviously be present in addition to the chemical role.

A complete discussion of these "rules", with the help of various examples has been developed.[10] A few supplementary examples of reactions belonging to the different classes are given here.

Some ionic reactions occurring in homogeneous solutions have been shown to remain unaffected by sonication, among which are acid catalysed ester hydrolysis and acetalisation, and aromatic deprotonation.[11] On the other hand, many more examples have been found for the promotion of electron transfer processes. In the typical example of the Kornblum-Russell reaction, sonication enhances the $S_{RN}1$ process.[12] Ando *et al.* recently published another illustrative example, that of the reaction of lead tetraacetate with styrene, shown in Figure 2.[13]

Type 2 heterogeneous ionic reactions should be easier to find. However the limit between "purely ionic" and "purely radical" processes can be difficult to determine in a number of examples. The acid catalysed sugar acetalisation seems to be one case without doubt. The method described for the synthesis of isopropylidene and cyclohexylidene derivatives[14] has been extended to benzylidene compounds, achieved in the presence of zinc chloride as the catalyst and is shown in Figure 3.[15]

The last case, that of heterogeneous processes involving electron transfer steps, especially using metals, is an important and well developed area of sonochemistry. One reason for this is that the first rational approach of sonochemistry relied mainly on the mechanical effects of ultrasonic cavitation on metals, which are known to react with organic substrates via single electron transfers.[16] Recent examples are found with the following reactions.

O_2N–C_6H_4–CH_2Br + $[Me_2C{=}NO_2]^-$ Li^+

SN_2 (stirring) → O-Alkylation product: 60%

$S_{RN}1$)))) → C-Alkylation product: 62%

Ph–CH=CH$_2$ + $Pb(OAc)_4$

radical pathway / mixed pathway / ionic pathway

	radical	mixed	ionic
(stirring) 50°C, 1h	0%	0%	5%
)))), 50°C, 1h	38%	12%	3%

Figure 2: Two cases of sonochemical switching

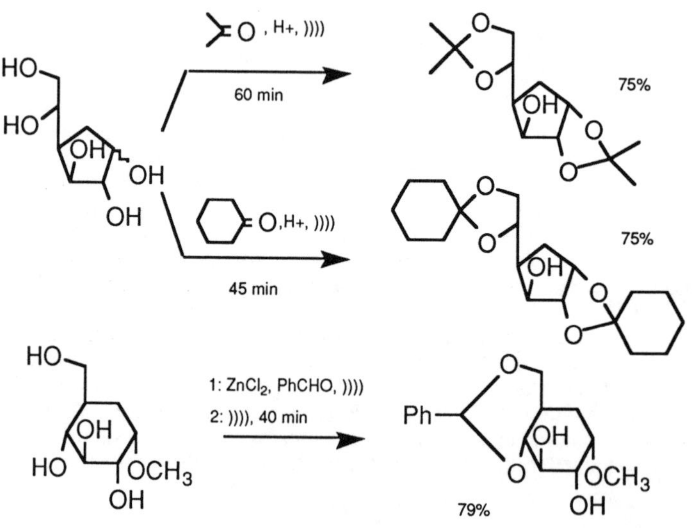

Figure 3: Sugar ketalisation.

The bromo iron complex shown in Figure 4, treated with ultrasonically dispersed potassium, yields the potassium derivative in excellent yield. This compound is alkylated under sonication to the ethyl complex, with a 80% yield for the two step sequence.[17] The second example reports the synthesis of several ruthenium-arene complexes from a single precursor, diaryl diruthenium tetrachloride. This compound is reduced by sonication in the presence of zinc to a probable unsaturated species, which is able to react with a variety of compounds to give mono and polynuclear complexes.[18] Some of these reactions have been carried out under triphasic conditions. Reagents such as gaseous ethylene and hydrogen have been used with success. Conditions have been found where ligand metal bonds can be reacted selectively (the Ru-Cl bond), while the others are left intact.

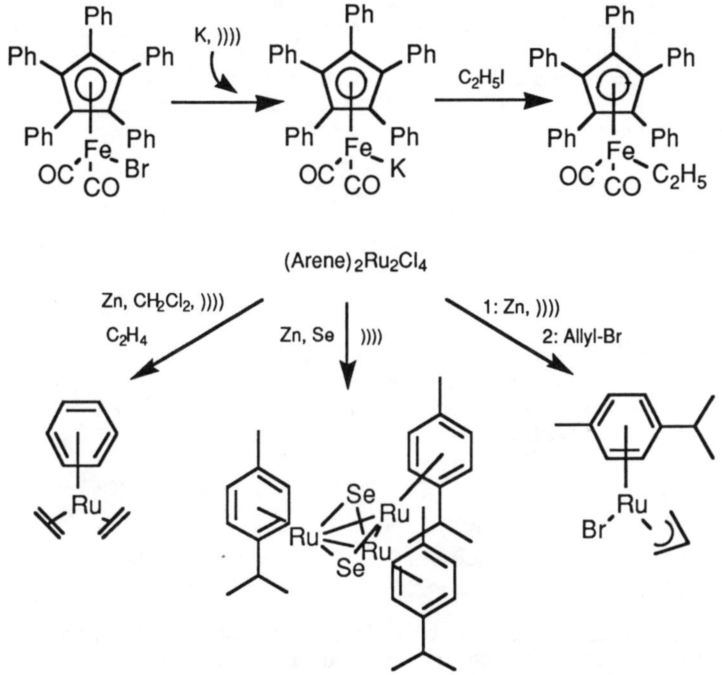

Figure 4: Sonochemical synthesis of transition metal complexes.

The empirical rules given above, together with most of the examples in the Literature, suggest that specific laws govern sonochemistry. If we consider that most of these chemical transformations are related to ultrasonic cavitation, the parameter "time" has to be taken into account. The reactive species, radicals, radical-ions, low coordination metal complexes, are generated on a very short time scale. Whether their origin is in the transitory high temperatures (hot spot theory[19]), or in electrical phenomena,[20] they undergo an extremely rapid, *ca.* 10^{-6} s, quenching to room temperature. In this manner, sonochemistry is clearly distinct from thermochemistry or flash thermolysis, for which the time scale is generally three or four orders of magnitude larger.

3 SONOCHEMISTRY AS AN IMPORTANT TOOL FOR SYNTHESIS

It is possible to find many organic reactions which illustrate the synthetic possibilities offered by sonochemistry. What is of growing importance now is the inclusion of sonochemical steps in complex, multistep synthetic sequences.

Methodological studies

Among recent sonochemical synthetic methods we find the addition of *p*-tosylazide to electron rich olefins, followed by nitrogen extrusion and rearrangement as shown in Figure 5.[21] The authors mention that optimum yields are obtained in the absence of solvent and that the reaction proceeds faster in liquids with a lower viscosity. These observations support the cavitational mechanism. On one hand, propagation of the waves is easier in low viscosity media. On the other hand, the preferential cavitation of the solvent, when present, reduces the rate of nitrogen extrusion from a non volatile molecule which has to diffuse near the cavitation bubble before undergoing the reaction. The crucial sonochemical step can be the extrusion of nitrogen, for which a radical mechanism is possible as in the case of the sonolysis of diazonium salts.[22]

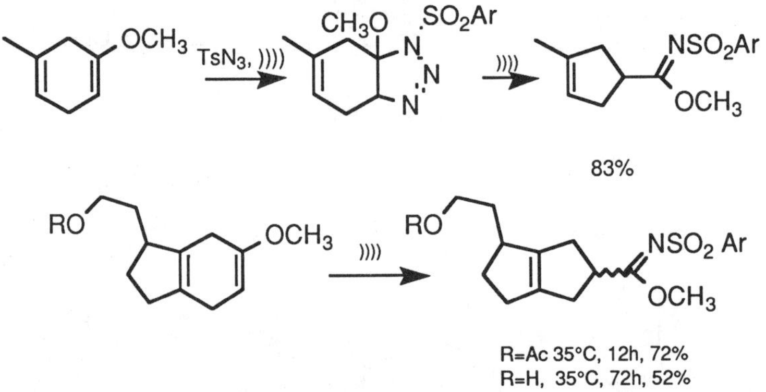

OCH_3 TsN_3,)))) CH_3O SO_2Ar N N N)))) NSO_2Ar OCH_3

83%

RO OCH_3)))) RO NSO_2 Ar OCH_3

R=Ac 35°C, 12h, 72%
R=H, 35°C, 72h, 52%

Figure 5: Cycloaddition-rearrangement of enol ethers.

Examples of sonochemical methods in organometallic synthesis are numerous. A recent case studied in our laboratory consists of the conjugate addition of functional alkyl halides to electron deficient olefins, using derivatives usually incompatible with classical organometallic methods.[23] A γ,δ-epoxyalkyl halide adds easily to various substrates in the presence of the zinc-copper couple in an aqueous solvent. The mechanism probably involves a one-electron reduction of the C-Halogen bond to give the corresponding nucleophilic radical. A carbanionic intermediate should be discounted, since hydrolysis is a minor reaction and no intramolecular attack on the epoxide ring is observed. Figure 6 gives an illustration of this method. With α,β-epoxyalkyl halides, the mechanism is quite different and no conjugate addition is obtained. The 3-membered ring is cleaved to an allylic alcohol in high yield and selectivity.[24] The intermediate case of β,γ-derivatives gives complex results which could not be made selective. Alcohols with a cyclopropyl methanol structure are formed, but the yields remain in the 30-60% range.[23]

A sonochemical use of Raney Nickel is shown in the following example. This catalyst, modified with tartaric acid, is treated with sodium bromide to deactivate the non-enantioselective sites.[25] The preparation of the catalyst, and the reduction of β-diketones or β-ketoesters are both effected under sonication. Under these conditions, the rates, yields and selectivities are substantially improved as shown in Figure 7.

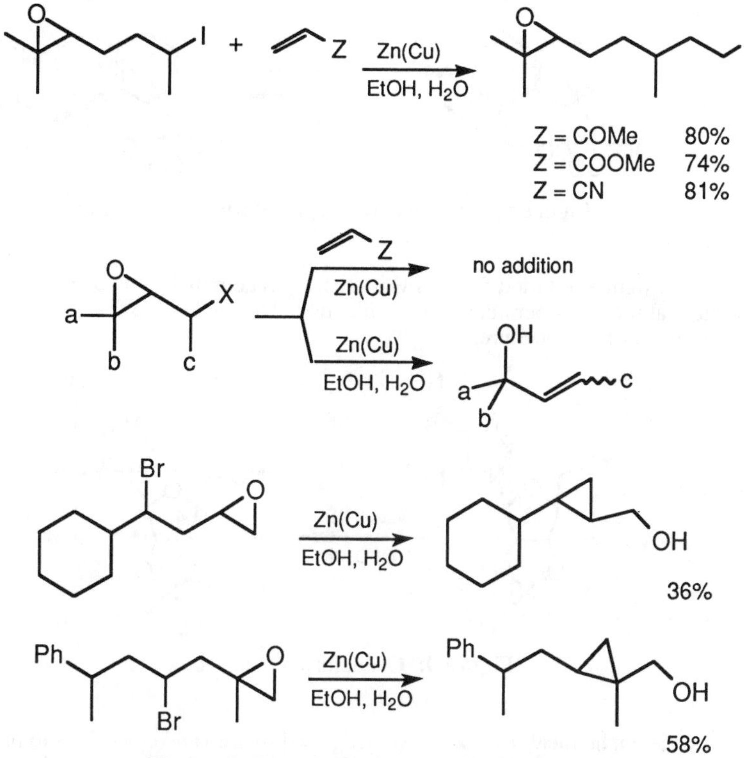

Figure 6: Reactions of epoxy alkyl halides.

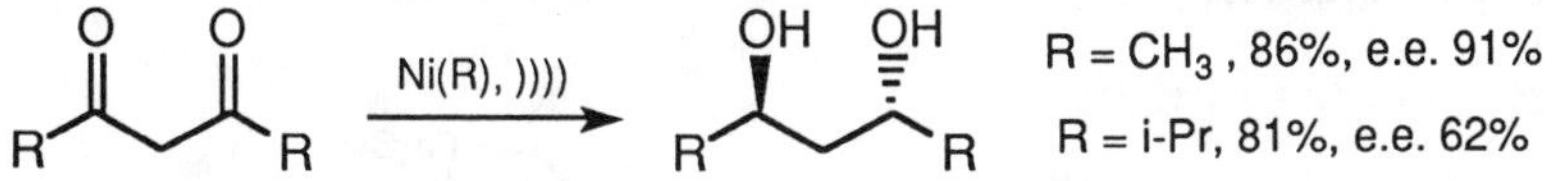

Figure 7: Enantioselective reductions with Raney Nickel.

Multistep synthesis

Methodological studies are generally achieved using simple molecules as models, which makes the problem of selectivity of secondary importance. The situation changes as soon as a given method has to be used during a multistep synthesis, especially when a complex (and expensive) molecule is the substrate. It is then necessary that the physical agent, the sonic waves, induces the desired reactions with minimal side effects. The following examples will illustrate these aspects. In Figure 8, the trifluoro acetamide can be hydrolysed in the presence of potassium carbonate in aqueous methanol.[26] The selectivity seems to be convenient as the authors do not mention a loss of the chromium tricarbonyl group.

A second example is found with the synthesis of a modified nucleoside, given in Figure 9.[27] Introduction of a cyano group in the 5' position requires displacement of the corresponding tosylate with potassium cyanide and 18-crown-6 in dioxane, or sodium cyanide in DMSO. Both methods suffer from various draw-

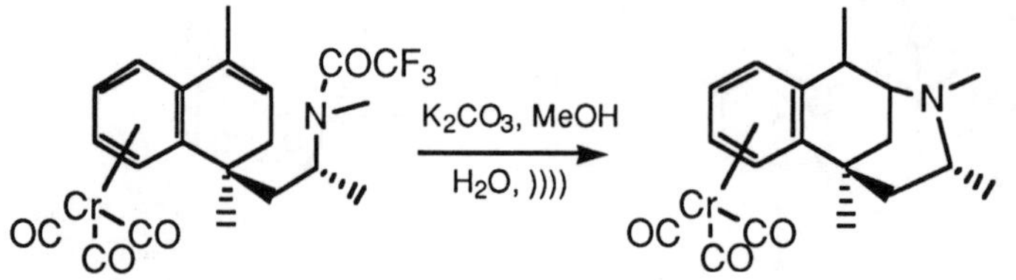

Figure 8: Selective hydrolysis of trifluoroacetamide.

backs. A significant modification was recently accomplished by using the second reagent at room temperature under sonication. The nucleoside framework is respected in this procedure.

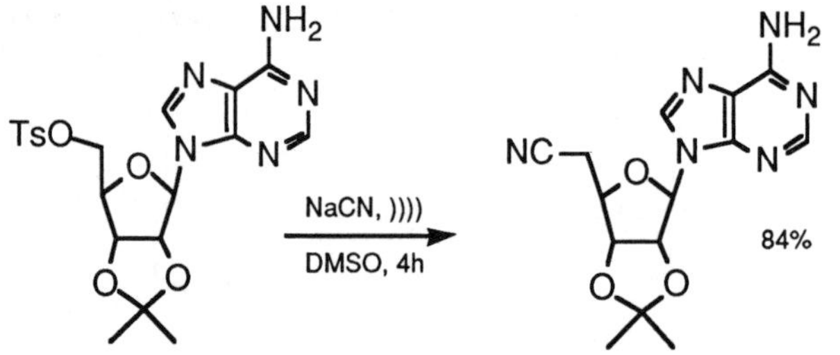

Figure 9: Cyanonucleoside synthesis.

Sonochemical methods have been used by Ley and coworkers to prepare the highly versatile ferrilactones, from vinyl epoxides and di-iron nonacarbonyl[28] as in Figure 10. The sonochemical reaction can be run with equal success either in early steps with structurally simple molecules, or with complex compounds.

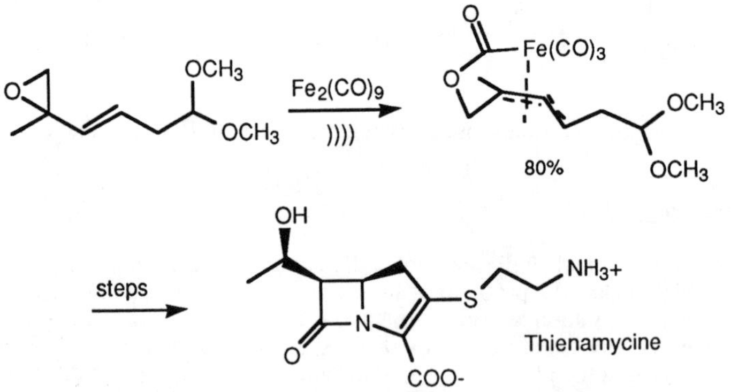

Figure 10: Ferrilactones in synthesis.

Some organometallic steps have been used in the total synthesis of natural compounds. The Barbier procedure[29] has been applied in several syntheses. The synthesis of Gascardic acid starts with the addition of an ethyl group to the carbonyl of 3-ethoxy-2-methyl-2-cyclopentene-1-one,with an excellent 98% yield as in Figure 11.[30] Similar reactions have been described by Mehta *et al.*,[31] with probable yield advantages over the silent process, although no comparative figures are given.

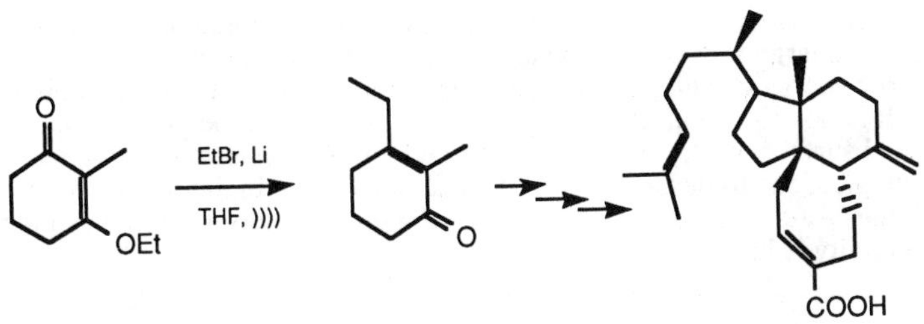

O EtBr, Li THF,)))) OEt O COOH

Figure 11: The initial step in the synthesis of Gascardic acid.

Zinc promoted reactions such as those in Figure 12 have been used in recent work. As an illustration of our epoxide ring opening method,[24] α- and β-Damascones, two appreciated odoriferous molecules, have been synthesised in good yields.[32] A new annelation method was developed for the synthesis of Corynantheine alkaloids, with an intramolecular Blaise reaction. This procedure can also be applied to more simple systems and a few examples are given.[33]

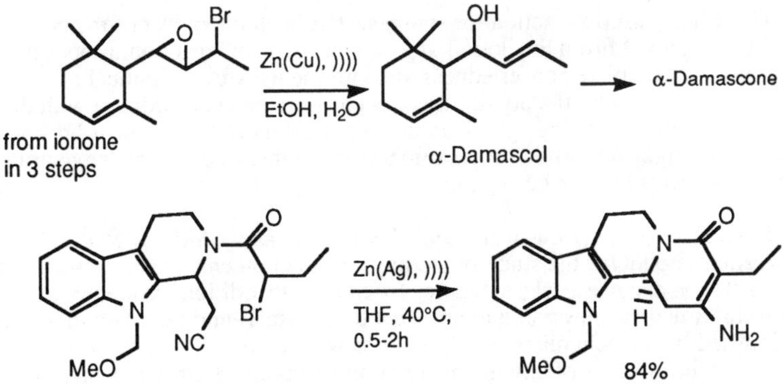

Br O Zn(Cu),)))) EtOH, H_2O OH → α-Damascone

from ionone in 3 steps α-Damascol

N O N Br NC MeO Zn(Ag),)))) THF, 40°C, 0.5-2h N O N H NH_2 MeO 84%

Figure 12: Zinc promoted sonochemical reaction.

The last example given in this part is illustrative of the importance of the irradiation conditions. The Vindolinine ring system shown in Figure 13 can be efficiently constructed from 19-iodo Tabersonine.[34] The authors effected this cylisation in the presence of sodium under high energy sonication with a probe.

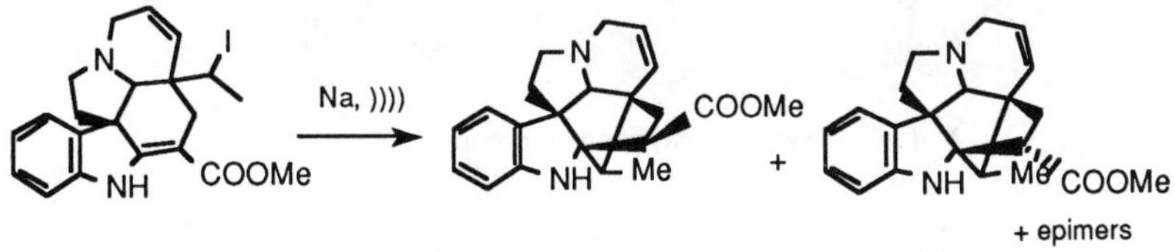

N I NH COOMe Na,)))) N COOMe NH Me + N NH Me COOMe + epimers

Probe, high energy	14%	50%
Probe, low energy	33%	66%
Cleaning bath		100%

Figure 13: Synthesis of Vindolinine.

They obtained a satisfactory yield of a mixture of the four possible diastereomers. Reducing the energy proved to be beneficial and excellent selectivity is obtained in a simple low energy cleaning bath. Under these conditions, only one stereoisomer is obtained. It is possible that, using low energy irradiation, all the process takes place on the metal surface, where the degrees of freedom of the intermediate are low. Higher energies lead to desorption and the reaction occurs in the liquid phase, where all the relative orientations of the reacting groups become possible, and the stereoselectivity is lower.

4 SONOCHEMISTRY AS A TOOL FOR THE STUDY OF REACTION MECHANISM

Since it has been established that the reactions which respond to sonochemical irradiation should proceed, at least in part, *via* single electron transfer mechanisms, it seems logical to use sonochemistry as an investigative method to give information on complex reaction pathways. There have not been many studies published until now, but two examples can be considered as demonstrative.

The Diels Alder reaction

This highly useful reaction for synthesis has been the object of many mechanistic studies.[35] From the notion of a "no mechanism" reaction, proposals have evolved to complete concertedness, the intermediacy of a biradical or a zwitterion. More recently, the possibility of catalysing the cycloaddition with the help of Lewis acids, oxidising agents (radical-cationic mechanism), or solid supports has brought substantial improvements in synthesis, but knowledge of its fundamental aspects has not been simplified.

From the point of view of the sonochemist, the Diels Alder reaction should be an excellent model for the study of the physical phenomena usually associated with cavitation as the classical procedures for this cycloaddition frequently use heat and pressure. It is then, surprising to note that only a small number of studies have been published by sonochemists. Two mentions were made of the lack of success of sonication, or the absence of any interesting improvements.[36] Only recently, in a series of papers, Snyder *et al.* reported successful results in the case of an electron rich diene adding to various ortho-quinones as in Figure 14.[37]

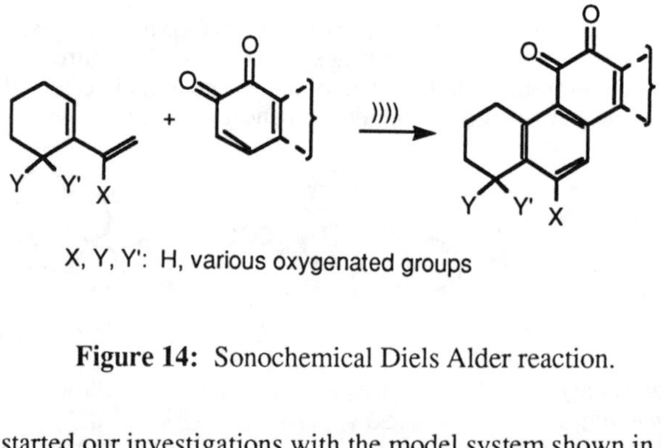

X, Y, Y': H, various oxygenated groups

Figure 14: Sonochemical Diels Alder reaction.

We started our investigations with the model system shown in Figure 15. Quinolinedione in toluene solution with an azadiene yields the expected 4+2 adduct in 63% yield under stirring and 70% under sonication.[38] This improvement seems modest, but more spectacular results are observed with the same reaction applied to naphthoquinones. Yields are not substantially modified, but the rate is increased. This ultrasound sensitive reaction should then, according to our "laws", proceed *via* an electron transfer step. In the presence of an oxidising catalyst, we were surprised to note that the yield undergoes a significant decrease. As the catalyst has no effect on the quinone, but oxidises the diene, we can conclude that the radical cations derived from the diene or the dienophile are not involved in the reaction mechanism. No EPR signal is observed when the azadiene is adsorbed on alumina, but a strong signal appears with the quinone on the same solid support. Addition to the latter system of the diene does induce the cycloaddition. The conclusion should be then that, in the case of the system studied here, the possibility exists for a new mechanism, in which the diene-dienophile couple first undergoes a redox reaction, followed by the addition of the quinone radical anion to the azadiene. It would be hazardous to extrapolate such conclusions to more general cases of Diels Alder reaction, but the results of the sonoreaction have been decisive in the formulation of the proposal of a new mechanism. Which are other favorable cases, and can we interpret the success of Snyder's experiments in a similar manner, are obviously important questions to be studied in the next stage of work.

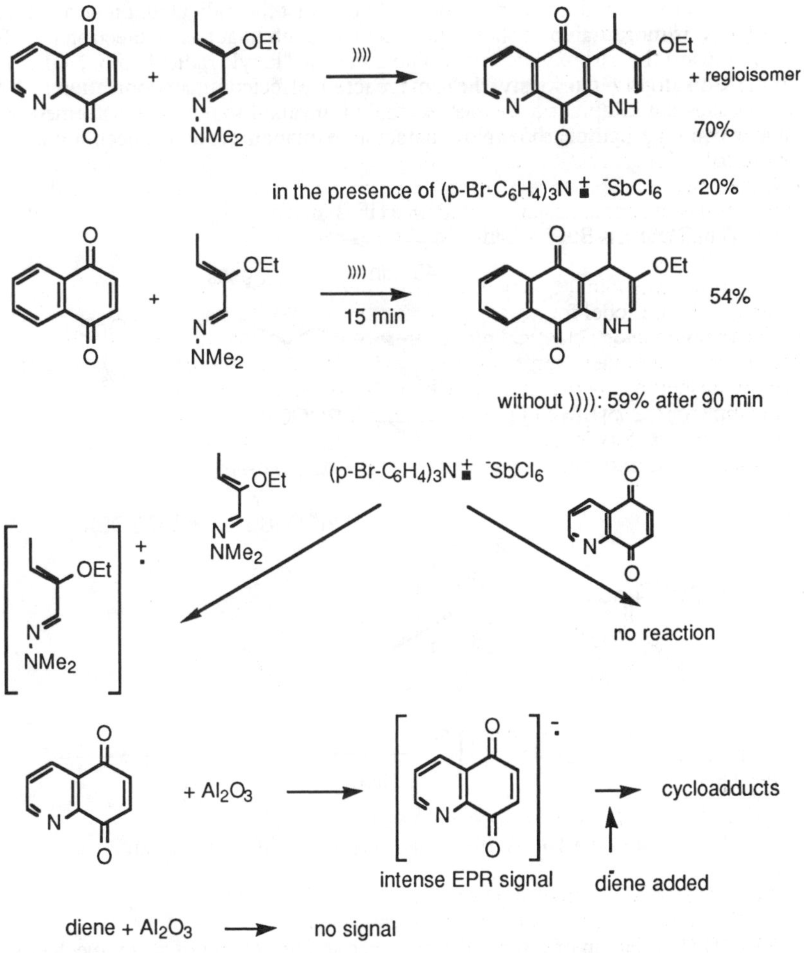

Figure 15: Hetero Diels Alder sonoreaction.

The Barbier reaction and some deductions with respect to the formation of organometallic compounds.

In continuation of our study of the sonochemical Barbier reaction, we became interested in its application to carboxylic acid derivatives. Addition of an alkyl group to the COO functionality is known to yield ketones. It would constitute a good synthetic method if the reaction were not frequently impaired by the undesired formation of tertiary alcohols. In an attempt to improve the procedure, we used the Barbier reaction with lithium salts of carboxylic acids.[39] It was then observed that the nature of the halogen in the alkyl halide was of major importance. Starting from the alkyl chlorides shown in Figure 16, sonication of the reaction mixture leads to high yields of the expected ketone.

On the other hand, surprising results were recorded with the iodide. In the presence of lithium benzoate, a 40% yield of benzil is obtained and no ketone is detected. In agreement with previous results,[40] we consider that the transformation of alkyl chlorides to the organolithium reagent is a fast process, essentially because of the high reactivity of the intermediate radical anion. After the alkyl lithium is formed it reacts selectively with the carboxylate.[41] In contrast the radical anion derived from the iodide is more stable. It should have time enough to act as an electron transfer agent toward the carboxylate salt, which gives the "ketyl" radical anion, a precursor to the α-diketone.[42] Curiously, the same reaction effected under sonication at 400 kHz leaves the salt practically unaffected, but the alkyl iodide is transformed to the Wurtz coupling hydrocarbon. No satisfactory explanation has yet been found for this result.

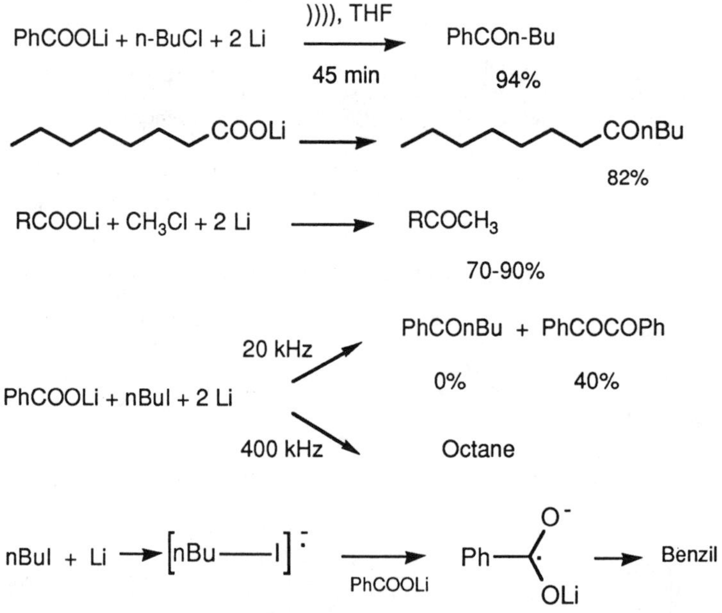

$$PhCOOLi + n\text{-}BuCl + 2\ Li \xrightarrow[45\ min]{)))),\ THF} PhCOn\text{-}Bu\ (94\%)$$

$$\text{(chain)COOLi} \longrightarrow \text{(chain)COnBu}\ (82\%)$$

$$RCOOLi + CH_3Cl + 2\ Li \longrightarrow RCOCH_3\ (70\text{-}90\%)$$

$$PhCOOLi + nBuI + 2\ Li \xrightarrow{20\ kHz} PhCOnBu\ (0\%) + PhCOCOPh\ (40\%)$$

$$PhCOOLi + nBuI + 2\ Li \xrightarrow{400\ kHz} Octane$$

$$nBuI + Li \longrightarrow [nBu\text{—}I]^{\cdot -} \xrightarrow[PhCOOLi]{} Ph\text{—}\dot{C}(O^-)(OLi) \longrightarrow Benzil$$

Figure 16: Barbier reactions on lithium carboxylate.

The formation of Grignard reagents is still the object of many mechanistic studies. A model recently developed is the D (for diffusion) model represented in

Figure 17.[43] After the initial electron transfer from the metal to the C-X bond, direct evolution to the organometallic is said not to be possible.[43,44] The radical anion is cleaved to the radical which then diffuses into the solution. It can there undergo various transformations: coupling, hydrogen abstraction from solvent, rearrangements, but these reactions are less probable than the re-adsorption onto the metal surface, where a second electron is transferred to give the carbanion. This complex representation of a reaction cascade leaves a question open, that of the necessity for the desorption-readsorption sequence. We envisaged that intra-molecular Barbier reactions could provide some new information on this point.

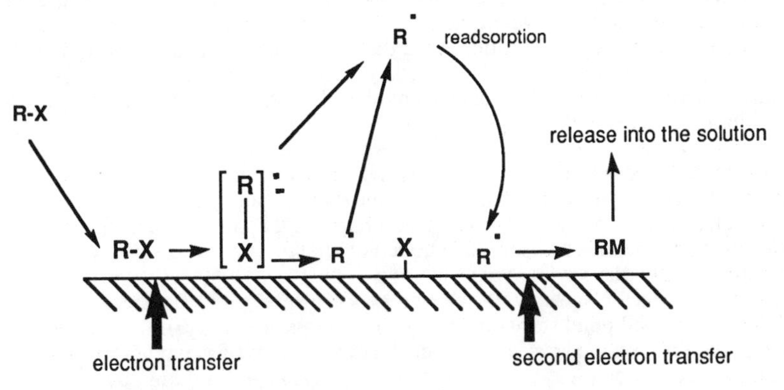

Figure 17: The D model for Grignard reagent formation.

When run with ω-halo-aldehydes or ketones, this reaction occurs with a variable efficiency, but it is reported to fail with ω-halo-esters and amides, leaving all the reactants unchanged.[45] It was generally felt that the absence of reactivity was due to physical passivation, a typical problem for which sonication can provide a solution. Using several salts and esters of ω-halogeno acids, we observed that sonication in the presence of lithium leads to interesting results demonstrated in Figure 18.

While the lithium or magnesium salts remain unreactive, the ethyl esters disappear in *ca.* 2 hours.[46] The yields of the expected ketone are still low, but the important result is that the attack does take place under sonication.

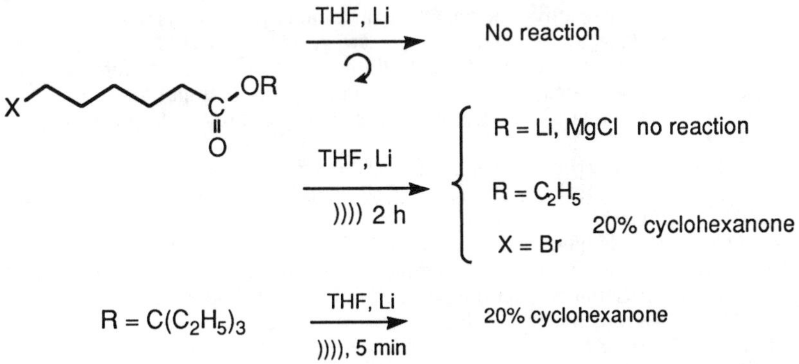

X OR C O

THF, Li → No reaction

THF, Li,)))) 2 h → R = Li, MgCl no reaction; R = C_2H_5, X = Br 20% cyclohexanone

R = $C(C_2H_5)_3$ THF, Li,)))), 5 min → 20% cyclohexanone

Figure 18: Intramolecular Barbier reaction of halo esters.

If we consider the D model, an interpretation can be proposed along the following lines. After the preliminary adsorption of the substrate onto the metal surface, the first electron transfer occurs. According to the D model this step should be followed by the desorption of the radical or the radical anion. Desorption of the radical anion, the Barbier reaction intermediate,[47] is possible only if it is accompanied by the extraction of the counter-cation from the metal lattice. The energy expense for this extraction is about 36 kcal/mole for magnesium and 37 kcal/mole for lithium, which are significant amounts of energy. Desorption of the radical, on the other hand, is possible after cleavage of the C-X bond if its interaction with the positively charged metal is not too strong. This is the case for simple alkyl groups. Negatively charged or electron rich groups (salts, amides, esters) should make this desorption more difficult, and the chemical evolution of the system goes to a dead end. Two solutions then exist, decreasing the interatomic bonding energy in the metal, or decreasing the adsorption energy of the reagent on the surface.

In the first case, sonication makes the extraction easier by producing fractures and lattice defects. In the bulk of the metal, magnesium and lithium have 12 neighbours. This number is only 9 and 8 respectively on a surface plane, 7 and 6 on an edge and 5 and 4 on a corner. The removal of an ion is obviously easier where the number of neighbours is lower.

The second point is that reducing the adsorption energy of the ester group can be made by using a sterically hindered group. 3-Ethyl-3-pentyl 6-bromo-hexanoate reacts completely in 5 min under sonication with lithium, but also in 40 min under stirring.[48] Even if the yield of cyclohexanone is low, this result demonstrates that the usual interpretation of the physical inhibition of the process reflects in part the reality, but chemical phenomena also contribute to this lack of reactivity. The D model is consistent with these results, which also explains the desorption-readsorption mechanism by the necessity for the radical to be re-adsorbed on an activated site where the extraction of an ion is less expensive in terms of energy.

5 SONOCHEMISTRY AND OTHER UNUSUAL DOMAINS OF CHEMISTRY

The usual interpretation of sonochemical phenomena is based on cavitation. To go further it would be necessary to understand the relative importance of thermal, electrical and/or mechanical effects. Many of these effects are compatible with the formation of highly reactive species, the exact nature of which is not clear. Flash thermolysis produces radicals. Ion radicals are generated by cold plasmas.[49] A comparison of sonochemical reactions with thermal or plasma reactions can be considered to provide guidelines for consideration, but the importance of "mechanical" chemical effects seems to have been generally underestimated.

In papers reporting on heterogeneous reactions one can read frequently the expression "simple mechanical effects". The word "simple" seems inappropriate, as it has been shown that the application of mechanical forces to a chemical system can cause chemical changes. This is an important point, for example in galenic pharmacy, as it is highly desirable that the transformation of a powder to a tablet does not induce chemical reactions. In fact, milling or compression produces radicals from various materials. Radical polymerisation occurs during ball milling of monomeric styrene, induced by electrons emitted by the shocks of balls on the metallic walls of the vessel.[50] In a general manner, mechanical forces applied to a solid lead to excited states which relax by emission of electrons.[51] In some cases EPR studies have been made. In crystalline solids, pulsed mechanical stresses generate structurally disordered, non-equilibrium states, which dissipate their energy excess by an enhanced chemical reactivity. An example is found with the mechanical decomposition of aluminum hydride, in which, as shown by Figure 19, the frequency of the stresses determines the course of the reaction.

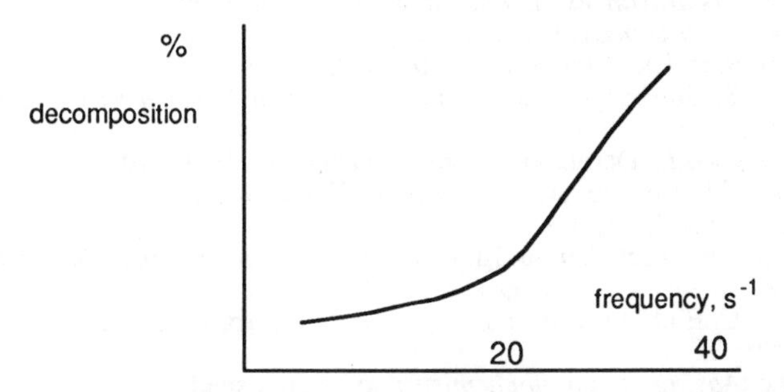

Figure 19: Decomposition of AlH_3 under the influence of mechanical shocks.

Pulsed mechanical stresses are precisely those undergone by a solid under ultrasound irradiation. It should then be of interest to determine if mechano-chemistry can be taken as a simplified model for the mechanical effects of sonochemistry.

6 CONCLUSION

Sonochemists interested in the fundamental phenomena which occur during irradiation are still facing important problems. One of these is the origin of sonochemistry. More and more observations give importance to electrical effects. However, shifting the interpretation from the hot spot theory to the electrical theory would probably be an exaggeration, and we think that the ideas should be complementary rather than exclusive. The superposition of thermal and electrical effects should be considered, even if the study of each component and the determination of its relative weight raises experimental problems. From the origins to the applications, the use of sonochemical reactions for synthetic purposes now finds increased interest, even from non-specialist chemists. The next step is obviously the scale-up to industrial processes.

If sonication can be considered as a new tool for the study of reaction mechanisms, many questions will arise. To illustrate this point, we will just cite one example. Olah *et al.* have described the strong acid catalysed isomerisation of polycyclic hydrocarbons to diamondoid cage compounds, under sonication.[52] They mention that the system is heterogeneous, so that the easiest interpretation would be that this reaction is a simple class 2 sonochemical process. However the question is not so easy to resolve and the presence of a free radical mechanism, or catalysis deserves consideration. Commeyras *et al.* have precisely investigated the rearrangement of branched alkanes in strongly acidic media (without sonication), and found that initiation can be made with free radicals.[53] Is such a process present in Olah's case? This would probably be an interesting topic for sonochemists. More generally, the meaning of the classical synchronous two electron transfer can be reexamined, and we hope that sonochemistry will provide new methods for a deeper knowledge of these fundamental ideas in organic chemistry.

REFERENCES

1. I. Elpiner *Ultrasound: Physical, Chemical and Biological Effects* Consultants Bureau, New York, N.Y. 1964.
2. M.A. Margulis *Advances in Sonochemistry* 1990, **1** 36.
3. A. Henglein *Ultrasonics* 1987, **25** 6.
4. P. Renaud *Bull. Soc. Chim. Fr.* 1950, 1044.
5. A.J. Fry, G.S. Ginsburg and R.A. Parente *J. Chem. Soc. Chem. Commun.,* 1978, 1040.
6. J.L. Luche and J.C. Damiano *J. Amer. Chem. Soc.* 1980, **102** 7926.
7. T.J. Mason, J.P. Lorimer and B.P. Mistry, *Tetrahedron Lett.* 1982, **23** 5563.
8. K.S. Suslick, P.F. Schubert and J.W. Goodale, *J. Amer. Chem. Soc.* 1981, **105** 7342.
9. J.L. Luche, C. Einhorn, J. Einhorn and J.V. Sinisterra-Gago *Tetrahedron Lett.,* 1990, **31** 4125.
10. J.L. Luche, *Advances in Sonochemistry* to be published.
11. J. Einhorn, C. Einhorn, M.J. Dickens and J.L. Luche *Tetrahedron Lett.,* 1990, **31** 4129.
12. M.J. Dickens and J.L. Luche *Tetrahedron Lett.*, 1991, **32** 4709.
13. T. Ando, P. Bauchat, A.Foucaud, M. Fujita, T. Kimura, H. Sohmiya *Tetrahedron Lett.* 1991 **32** 6379.
14. C. Einhorn and J.L. Luche *Carbohydr. Res.*, 1986, **155** 258.
15. G.J.F. Chittenden *Rec. Trav. Chim. Pays-Bas.,* 1988, **107** 607.
16. E.C. Ashby and J. Oswald *J. Org. Chem.*, 1988, **53** 6068; A. Samat, B. Vacher, M. Chanon *J. Org. Chem.*, 1991, **56** 3524.
17. P. Bregaint, J.R. Hamon and C. Lapinte *J. Organomet.Chem.,* 1990, **398** C25.
18. R.S. Bates and A.H. Wright *J. Chem. Soc. Chem. Commun.,* 1990, 1129.
19. B.E. Noltingk and E.A. Neppiras *Proc. Phys. Soc.* 1950, **63B,** 674.
20. Ya. I. Frenkel *Russ. J. Phys. Chem.* 1940, **14,** 305.
21. D. Goldsmith and J.J. Soria *Tetrahedron Lett.,* 1991, **32** 2457.
22. D. Rehorek and E.G. Janzen *J. Prakt. Chem.* 1984, **326** 935.
23. L.A. Sarandeses, A. Mourino, J.L.Luche *J. Chem. Soc. Chem. Commun.* 1992 in press.
24. L.A. Sarandeses, A. Mourino and J.L. Luche *J. Chem. Soc. Chem. Commun.* 1991, 818.
25. A. Tai, T. Kikukawa, T. Sugimura, Y. Inone, T. Osawa and S. Fugii *J. Chem. Soc. Chem. Commun.*, 1991, 795.
26. M. Sainsbury, C.S. Williams, A. Naylor and D.I.C. Scopes *Tetrahedron Lett.*, 1990, **31** 2765.
27. A.R. Singh, *Synth. Commun.* 1990, **20,** 3547.
28. S.V. Ley and C.M.R. Low *Ultrasound in Synthesis* Springer Verlag, Berlin, 1989, p 105.
29. J.C. de Souza Barboza, C. Petrier and J.L. Luche *J.Org. Chem.*, 1988, **53** 1212.
30. G. Berubi, A.G. Fallis *Tetrahedron Lett.,* 1989, **30** 1989.
31. G. Mehta and N. Krishnamurthy *J. Chem. Soc .Chem. Commun.,* 1986, 1319.
32. L.A. Sarandeses and J.L.Luche *J. Org.Chem.,* 1992, **57** in press.
33. R.L. Beard and A.I. Meyers *J. Org. Chem.,* 1991, **56** 2091.
34. G. Hugel, D. Cartier and J. Levy *Tetrahedron Lett.*, 1989, **30** 4513.
35. F. Fringuelli and A. Taticchi *Dienes in Diels Alder reactions,* J. Wiley, New York, 1990.
36. J. Elguero, P. Goya, J.A. Paez, C. Cativiela and J.A. Majoral *Synth. Commun.,* 1989, **19** 473; P.A. Grieco, P. Garner and Z. Ha *Tetrahedron Lett.* 1983, **24** 1897.

37. J. Lee and J.K. Snyder *J. Org. Chem.,* 1990, **55** 4995; M. Hainza, J. Lee and J.K. Snyder *J. Org. Chem.,* 1990, **55** 5008; J. Lee, H.S. Mei and J.K. Snyder *J. Org. Chem.* 1990, **55** 5013.
38. P. Nebois, H. Fillion, J.L.Luche to be published.
39. Danhui Yang, C. Einhorn and J.L. Luche unpublished results
40. J.C. de Souza Barboza, J.L. Luche and C. Petrier *Tetrahedron Lett.,* 1987, **28** 2013.
41. C. Einhorn, J. Einhorn and J.L. Luche *Tetrahedron Lett.*, 1991, **32** 2771.
42. R. Karaman and J.L. Fry, *Tetrahedron Lett.* 1989 **30** 6267.
43. J.F. Garst, F. Ungvary, R. Batlaw and K.E. Lawrence *J. Amer. Chem. Soc.,* 1991, **113** 5393.
44. A. Dubois and S. Nuzzo *J. Amer. Chem. Soc.* 1986, **108** 2881.
45. Y. Leroux *Bull. Soc. Chim. Fr.* 1968, 359.
46. J. Einhorn and J.L. Luche unpublished results.
47. A. Moyano, M. Pericas, A. Riera, J.L.Luche *Tetrahedron Lett.,* 1991, **31** 7619.
48. J.L. Luche, unpublished results.
49. H. Suhr and U. Kürzel *Liebigs Ann. Chem.* 1979, 2057.
50. C.V. Oprea and F. Weiner, *Ang. Makromol. Chem.*, 1984 **126** 89.
51. V.V. Boldyrev, N.Z. Lyakhov, Yu.T. Pavlyukhin, E.V. Boldyreva, E.Yu. Ivanov and E.G. Avvakumov *Mechanochemistry*, Soviet Scientific Rev., Harwood Acad. Press, London, 1990, Vol. 14, 105.
52. O. Farooq, M.F. Farnia, M. Stephenson and G.A. Olah *J. Org. Chem.* 1988, **53** 2840.
53. H. Choukroun, A. Germain, D. Brunel, A. Commeyras *Nouv. J. Chim.* 1983, **7** 83.

Ultrasound in Synthesis: an Overview

R. Bowser and R.S. Davidson
THE CHEMICAL LABORATORY, UNIVERSITY OF KENT, CANTERBURY CT2 7NH, UK

Undoubtedly the rapid progress that is being made in the fields of information retrieval and computer assisted planning of synthesis is playing an important part in increasing the efficiency with which target compounds can be synthesised. For these systems to be of value there needs to be a good bank of well tried synthetic methodologies for which, in the main, there is also an understanding of the mechanism of the reactions involved. Ultrasound offers a way of energising a chemical system along with the more conventional methods such as thermal, photochemical and electrochemical. If ultrasound is to have any impact upon synthetic organic chemistry it must open up new synthetic routes, offer a more expedient way of carrying out well documented reactions, improve regio and stereo selectivity of reactions, increase the ability to manipulate selected functional groups without affecting other groups and, ideally, enhance our ability to achieve enantioselectivity. In some cases, the use of ultrasound has been of value in these areas, but currently the predictability, *i.e.* a knowledge as to when its use will be valuable and the course of the reaction will be known with certainty, is somewhat lacking. One can say that the application of ultrasound will often increase the rate of heterogeneous reactions (by aiding mass transfer) and reactions involving metals (by cleaning and activating their surfaces). One can now add to this list, as a result of the work of Luche and colleagues[1], that single electron transfer processes (SET reactions) will be aided by ultrasound. Of particular significance is that for reactions where the reagents may react via either an ionic or free radical route, application of ultrasound facilitates the free radical at the expense of the ionic process. Three examples which illustrate this rationale are:

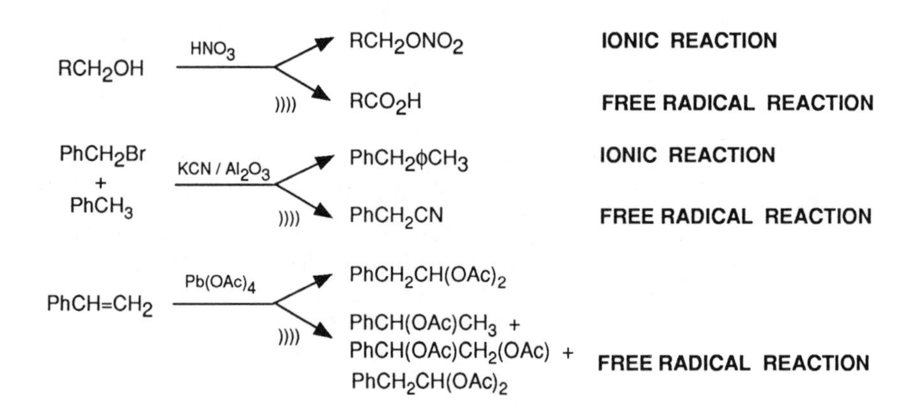

(i) the nitration of primary alcohols[2], (ii) competition between Friedel Crafts reaction and a nucleophilic substitution reaction[3] and (iii) the reaction of styrene and its derivatives with lead tetraacetate[4].

These examples lend credence to the theory put forward by Luche and colleagues and they are further substantiated by the results obtained from a study of $S_{RN}1$ reactions run with and without the aid of ultrasound. The reaction of the lithium salt of 2-nitropropane with 4-nitrobenzyl bromide which normally gives, in the main, products derived by O-alkylation, yields under the influence of ultrasound those derived by an $S_{RN}1$ mechanism. These findings must surely be the forerunners of many more which will put the mechanistic understanding and synthetic applicability of one aspect of ultrasound upon a firm basis.

An area of synthesis in which ultrasound has already made significant contributions is that in which metals are used as reagents. The usefulness of ultrasound to disperse lithium[6], sodium[7] and potassium[8] has been extensively exploited. Whilst this advantage, imparted by the use of ultrasound, is of enormous importance, the effect of ultrasound clearly goes further since it has been found that under certain reaction conditions the optically active halide (1) on lithiation gives a species which reacts with a ketone to give a tertiary alcohol with a high degree of retention of configuration at the chiral atom[9]. This reaction has been the subject of theoretical studies in which attempts were made to elucidate at which stage electron transfer occurs and the degree of association between anions and cations in the various intermediates[10]. These studies show, once again, that ultrasound has a profound, and as yet unexplained effect, upon electron transfer reactions.

n C_6H_{13}–C(CH_3)(H)(Hal) (1) —Li, US→ [n C_6H_{13}–C(CH_3)(H)(Hal)$^{\bullet -}$ Li^+] → n C_6H_{13}–$C^{\bullet}$(CH_3)(H) —Li→ n C_6H_{13}–C(CH_3)(H)(Li) —(CH$_3$)$_2$C=O→

Product with retention of configuration

A reaction, involving lithium which appears to be unprecedented is that of the reduction of carboxylic acids to give aromatic 1,2-diketones[11]. If this reaction can be scaled up and proves to be reliable then it would offer another route to these compounds which are normally prepared via the benzoin condensation and oxidation. Lithium and potassium, unlike sodium, can under the influence of ultrasound reduce triarylmethanols to triarylmethanes[12].

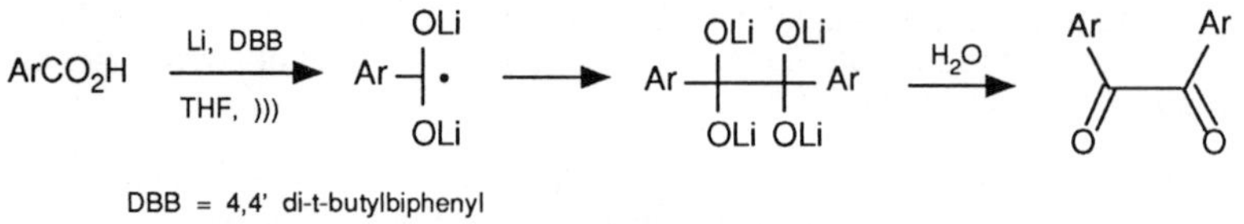

$ArCO_2H \xrightarrow[\text{THF,)))}]{\text{Li, DBB}} Ar-C^{\bullet}(OLi)_2 \longrightarrow Ar-C(OLi)_2-C(OLi)_2-Ar \xrightarrow{H_2O} ArCO-COAr$

DBB = 4,4' di-t-butylbiphenyl

Reductive deoxygenation of ketones to give alkenes by low valent titanium compounds is a well known and useful reaction. Application of ultrasound allows these reactions to be conducted under milder conditions[13]. Interestingly there is a marked solvent effect and use of dimethoxyethane leads to 1,2-diols isolated instead of the usual alkenes.

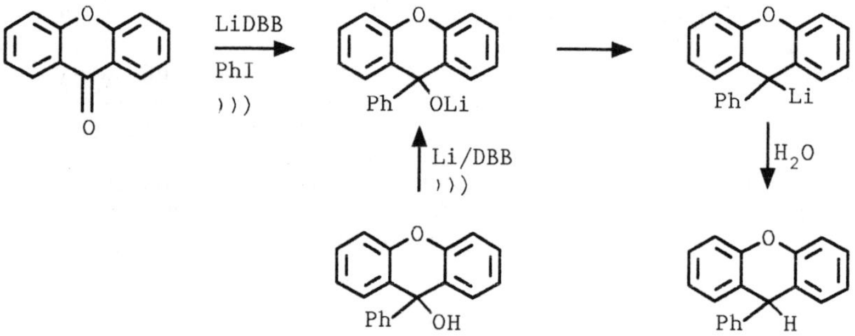

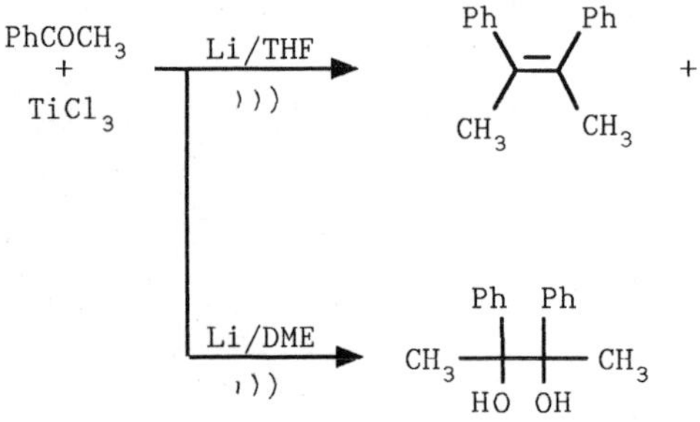

LiDBB, PhI,))) ; Ph, OLi ; Ph, Li ; Li/DBB))) ; H_2O ; Ph, OH ; Ph, H

$PhCOCH_3$ + $TiCl_3$; Li/THF))) ; Ph, Ph, CH_3, CH_3 + Ph, CH_3, CH_3, Ph ; Li/DME))) ; Ph Ph, CH_3, CH_3, HO OH

dl and meso

The potassium induced cleavage of a C-S bond in cyclic sulphones has proved to be of some mechanistic interest[14]. It has been suggested that radical anions are intermediates in these reactions[15]. The reactions have now been studied using d_8-toluene as solvent[16] and the products found to contain deuterium. The deuterium is believed to be incorporated *via* deuteron abstraction from toluene. In accordance with this view, quenching the reaction mixture with deuterium oxide also leads to deuterium incorporation. The regiospecific incorporation of deuterium expands the synthetic utility of this reaction.

$$\text{sulfolane } (SO_2) \xrightarrow[\text{2. } D_2O,\ THF,\ CH_3I]{\text{1.))))\ K, THF}} SO_2^{-}\text{ radical} \xrightarrow{K} SO_2^{-}\text{ anion} \longrightarrow \text{D-substituted } SO_2CH_3$$

Perhaps one of the most familiar uses of ultrasound is to initiate the formation of Grignard reagents. Ultrasound can have a most dramatic effect upon metals which includes scouring away oxide layers (*via* microjects and streaming), eroding the surface thereby increasing the surface area of the metal, and even causing fusion of metal particles[17]. Observation of the latter is taken as evidence for the correctness of the "Hot Spot Theory". Activation of magnesium has enabled Grignard reactions to be carried out in wet solvents,[18] the preparation of allylic Grignard reagents[19] (something which is particularly difficult to do using other means), and to provide a synthetically useful way of preparing trialkylboranes from Grignard reagents[20]. Magnesium is also a powerful reducing agent and can reduce carbonyl groups, and under the influence of ultrasound, anthracene[19].

Allylstannanes are of particular synthetic utility[20] and are normally prepared by reaction of the appropriate metallated allyl halide with the appropriate stannyl halide. In many cases these reactions produce the products in low yield, and consequently, the finding that the reaction of allylic halides with chlorostannanes in the presence of magnesium and application of ultrasound gives the product in high yield and purity, is of synthetic importance[22], *e.g.*

$$PhCH{=}CHCH_2Cl + Bu_3SnCl \longrightarrow PhCH{=}CHCH_2SnBu_3$$

100% isolated yield

The authors reported that use of chlorotrimethylsilane in place of the chlorostannane did not give the expected allylsilanes. Given the importance of the latter in synthetic chemistry[23] we re-investigated the reactions. We found that benzylic halides would react with chlorotributylstannane and chlorotrimethylsilane in tetrahydrofuran solution containing magnesium, and under the influence of ultrasound to give high yields of the desired products in short reaction times. The method also proved useful for the preparation of the corresponding naphthyl compounds (Table 1). Although many of these compounds have been prepared by other routes, the described route has the advantage of simplicity, the products being produced in higher yield and purity, and under milder reaction conditions. Thus 2-naphthylmethyl tributylstannane can be prepared by reaction of 2-naphthylmethyl lithium with chlorotributylstannane in 57% yield[24]. Our described method obviates the need to prepare the lithiated naphthalene species.

In connection with another project we required some triorganosilylalkylamines. Various routes exist and have been described for the synthesis of these compounds, but most of them are multistep[25]. The most direct route for making the stannyl derivatives involves lithiation of tributyltin chloride at 0 °C, followed by reaction with an α-dimethylamino-ω-haloalkane[26]. Using this method, 3-(N,N-dimethyl-amino)propyl tri-*n*-butyltin was obtained in 18% yield. We have found that α-dimethylamino-ω-chloroalkanes react with trimethylsilyl chloride and tri-*n*-butyltin chloride to give reasonable yields of silylalkylamines and stannylalkylamines respectively in the presence of magnesium and with the application of ultrasound at room temperature. As can be seen from Table 2, the method is quite versatile and reaction times relatively short. The mechanism of the reaction is of interest since unlike the methodologies not involving ultrasound, there is no prior formation of the metallated alkyl halide.

Table 1: Yields of benzylic stannanes ans silanes obtained by reaction of benzylic halides with chlorotrimethylsilane and chloro-trimethylstannane.

COMPOUND	REACTION TIME (hour)	YIELD (%)
$CH_3-C_6H_4-CH_2SnBu_3$	1.0	80
$CH_3O-C_6H_4-CH_2SnBu_3$	3.5	100
$CH_3O-C_6H_4-CH_2Si(CH_3)$	3.5	98
1-naphthyl-$CH_2Si(CH_3)_3$	10.0	42
2-naphthyl-$CH_2Si(CH_3)_3$	10.0	88
1-naphthyl-CH_2SnBu_3	5.0	90
2-naphthyl-CH_2SnBu_3	5.0	78

Thus we cannot be sure if the sequence of events begins with metallation of the silyl/stannyl chloride or whether the metallated alkyl halide is formed and then reacts rapidly with the silyl/stannyl chloride. There is no doubt however, that the success of this synthetic strategy depends upon the activation of magnesium by ultrasound.

Another metal whose synthetic utility has been substantially increased via the application of ultrasound is zinc. Studies have been made of the effect of ultrasound upon zinc[17] and it was found that its application led to the removal of the surface zinc oxide layers, caused changes in particle size and caused smaller particles to fuse together. Some of the more notable findings include the generation of perfluoro-alkylzinc halides[27], facilitating the Reformatsky reaction[28] which includes an example where the reaction will not proceed without the application of ultrasound[29] and in the generation of reagents akin to organozincs[30]. The latter reagents undergo many of thereactions which are typical of alkylzincs but especially in the case of "allylzincs" there are indications that yet again ultrasound is promoting reaction via a single electron transfer process. Luche and colleagues have pioneered the application of reagents derived from zinc-copper couples, using ultrasound, to synthetic organic

Table 2: Silyl and stannyl alkylamines obtained *via* the route utilising ultrasound.

COMPOUND	REACTION TIME (hours)	YIELD (%)
$Me_2N(CH_2)_3SiMe_3$	4	37
$Me_2N(CH_2)_3SnBu_3$	2	35
$Me_2N(CH_2)_2SiPh_3$	4	40
$Me_2N(CH_2)_2SnPh_3$	4	35
$Me_2N(CH_2)_3SiPh_3$	4	30
$Me_2N(CH_2)_2SnMe_3$	2	45

chemistry[31]. Evidence has been produced which supports the view that these reactions also involve single electron transfer. The reactions based upon zinc-copper couples are finding extensive use, e.g. in a synthesis of vitamin D, [32] enantiomerically pure protected unsaturated α-amino acids[33] and in the ring opening of glycidyl halides to produce allylic alcohols[34]. The reducing power of zinc has been well exploited in organic chemistry but once again ultrasound affects the course of these reactions which undoubtedly involve a single electron process as is the case in the latter reaction. Ultrasound enables zinc and acetic acid to be used to reduce the carbon-carbon double bonds of unsaturated esters and quinones[35]. The reduction of diarylketones by zinc in the presence of aluminium trichloride leads to reductive coupling[36].

Ultrasound has found application in activating catalysts such as platinum[37], palladium[38] and nickel[39]. The activation of nickel by ultrasound has recently been shown to be of value in the hydrogenolysis of the N-N bond of hydrazines[40] and importantly, in catalysing the hydrosilylation of alkenes[41]. This latter result is particularly exciting as it may lead to a way of replacing the more usual expensive catalyst (platinum) used for this reaction by the less expensive nickel. A most intriguing finding has been reported in which Raney nickel modified by interaction with tartaric acid and sodium bromide is transformed, by the use of ultrasound, into an enantioselective catalyst[42]. The catalyst was shown to reduce β-ketoesters and β-diketones with high enantioselectivity. The rationale behind using ultrasound is that the non enantiodifferentiating sites on the nickel surface are characterised by exhibiting greater disorder than the enantioselective sites and hence ultrasound preferentially erodes away the non enantiodifferentiating sites. This result and the philosophy behind the work could be of great value in developing new processes and in improving known reactions. The reported results describe a method which may well effectively compete with enzymatic methods.

meso

60% isolated
optically pure

A less glamorous role for ultrasound in synthesis, albeit extremely valuable, is in increasing the effectiveness of mass transfer, i.e. mixing. We have shown that N-alkylation reactions, esterification and the hydrolysis of esters can benefit through the application of ultrasound[43]. Heterogeneous reactions are prime targets for this application and recent examples include the ring opening of oxiranes by hydrogen fluoride[44] and the addition of "IF" to alkenes[45].

KHF_2/AlF_3))) 49%

NIS, KHF_2/AlF_3)))

NIS= N-Iodosuccinimide

The reagent mix - potassium hydrogen difluoride - porous aluminium trifluoride appears to generate hydrogen fluoride. This is another example of a reaction which does not occur in the absence of ultrasound. A synthetically useful reaction which also benefits from the use of ultrasound is the Reimer-Tiemann reaction[46]. The ultrasound is used to generate dichlorocarbene in a biphasic system. Interestingly the use of probe was found to be more effective in aiding the reaction than an ultrasonic bath. Ultrasound has also found use in aiding the removal of silyl protecting groups (*t*-butyldimethylsilyl by fluoride ions)[47], in facilitating oxidation reactions employing pyridinium chlorochromate[48] and tetra-*n*-propylammonium perruthenate[49], and in some O- and C-alkylation reactions[50].

In conclusion, it should be pointed out that many homogeneous reactions also benefit from the application of ultrasound and the majority do not appear to involve single electron transfer processes. Examples of such reactions include the conversion of aryldiazonium salts to aryl ethers[51], the reaction of sulfonyl azides with alkenes[52], generation of the ketones[53] and ketiminium ions[54] facilitating esterification reactions[55] and the reduction of esters by zinc borohydride[56]. A useful aspect of the latter reaction was the finding that under appropriate conditions aliphatic esters are preferentially reduced with aromatic esters remaining unreacted.

To what extent has the challenge to ultrasound been met? Some new reactions have been found, the efficiency of many reactions has been improved and in the case of catalysts, an improvement in enantio-selectivity has been achieved. Our understanding of the mechanism of processes induced by ultrasound is increasing, but there are still many unresolved questions.

ACKNOWLEDGEMENTS

We should like to thank the SERC for a CASE award (to RB) in which Cookson Research were the industrial partners.

REFERENCES

1. J.-L. Luche, C. Einhorn, J. Einhorn and J.V. Sinisterra-Gago, *Tetrahedron Lett.* 1990, **31** 4125.
2. C. Einhorn, J. Einhorn, M.J. Dickens and J.-L. Luche, *Tetrahedron Lett.* 1990, **31** 4129.
3. T. Ando, T. Kawate, J. Ichihara and T. Hanafusa, *Chem. Lett.*, 1984, 725.
4. T. Ando, P. Bauchat, A. Foucaud, M. Fujita, T. Kimura and H. Sohmiya, *Tetrahedron Lett.* 1991, **32** 4709.
5. M. J. Dickens and J.-L. Luche, *Tetrahedron Lett.* 1991, **32** 4709.
6. P. Boudjouk, R. Sooriyakumaran and B.H. Han, *J. Org. Chem.*, 1986, **51** 2818; J.-L. Luche and J.C. Damiano, *J. Amer. Chem. Soc.*, 1980, **102** 7926; J. Einhorn, J.-L. Luche and P. Demerseman, *J. Chem. Soc. Chem. Commun.*, 1988, 1350; J. Einhorn and J.-L. Luche, *J. Org. Chem.*, 1987, **52** 4124.
7. M.W.T. Pratt and R. Helsby, *Nature (London)*, 1959, **184** 1694; S.V. Ley, I.A. O'Neil and C.M.R. Low, *Tetrahedron* 1986, **42** 5363; J. Einhorn and J.-L. Luche, *Tetrahedron Lett.* 1986, **27** 501.
8. J.-L. Luche, C. Petrier and C. Dupuy, *Tetrahedron Lett.*, 1984, **25** 753.
9. J.C. de Souza-Barboza, J.-L. Luche and C. Petrier, *Tetrahedron Lett.*, 1987, **28** 2013.
10. A. Moyano, M.A. Pericas, A. Riera and J.-L. Luche, *Tetrahedron Lett.*, 1990, **31** 7619.
11. R. Karaman and J.L. Fry, *Tetrahedron Lett.*, 1989, **30** 6267.
12. R. Karaman, D.T. Kohlman and J.L. Fry, *Tetrahedron Lett.*, 1990, **31** 6153.
13. S.K. Nayak and A. Banerji, *J. Org. Chem.*, 1991, **56** 1940.
14. T.-S. Chou and M.-L. You, *Tetrahedron Lett.*, 1985, **26** 4495.
15. R S Davidson in *Chemistry With Ultrasound*, T J Mason (Ed.), Society of Chemical Industry, Elsevier Applied Science, London, 1990, Ch. 3, 65.
16. T.-S. Chou, S.-H. Hung, M.-L. Peng and S.-J. Lee, *Tetrahedron Lett.*, 1991, **32** 3551.
17. K.S. Suslick and S.J. Doktycz, *Advances in Sonochemistry*, 1990 **1** 197; K.S. Suslick, *Science*, 1990, **247** 1373.
18. J.P. Sprich and G. Lewandos, *Inorg. Chim. Acta.*, 1983, **76** 241.

19. W. Oppolzer and P. Schneider, *Tetrahedron Lett.*, 1984, **25** 3305.
20. H.C. Brown and U.S. Racherla, *Tetrahedron Lett.*, 1985, **107** 1778.
21. B.M. Trost and P.J. Bonk, *J. Amer. Chem. Soc.*, 1985, **107** 1778; A.J. Pratt and E.J. Thomas, *J. Chem. Soc. Chem. Commun.*, 1982, 1115; Y. Yamamoto, H. Yatagai, Y. Naruta and K. Maruyama, *J. Amer. Chem. Soc.*, 1983, **48** 1558.
22. Y. Naruta, Y. Nishagaichi and K. Maruyama, *Chem. Lett.*, 1968, 1957.
23. A. Hosomi, *Acc. Chem. Res.*, 1988, **21** 200.
24. M.A. Tius and J. Gomez-Galeno, *Tetrahedron Lett.* 1986, **27** 2571.
25. Y. Sato, Y. Ban and H. Shirai, *J. Organomet. Chem.*, 1976, **113** 115.
26. J.W. Suggs and K.S. Lee, *J. Organomet. Chem.* 1986, **299** 297.
27. T. Kitazume and N. Ishikawa, *Chem. Lett.*, 1981, 1679; T. Kitazume and N. Ishikawa, *J. Amer. Chem. Soc.*, 1985, **107** 5186; A. Solladie-Cavello, D. Farkhani, S. Fritz, T. Lazrak and J. Suffert, *Tetrahedron Lett.* 1985, **25** 4117.
28. B.H. Han and P. Boudjouk, *J. Org. Chem.* 1982, **47** 5030; R.W. Land and B. Schaub, *Tetrahedron Lett.* 1988, **29** 2943.
29. T. Kitazume, *Synthesis,* 1986, 855.
30. C. Petrier and J.-L. Luche, *J. Org. Chem.* 1985, **50** 910.
31. C. Petrier, C. Dupuy and J.-L. Luche, *Tetrahedron Lett.*, 1986, **27** 3149; J.-L. Luche, A. Allavena, C. Petrier and C. Dupuy, *Tetrahedron Lett.*, 1988, **29** 5373.
32. J.L. Mascarenas, J. Perez-Sestelo, L. Castedo and A. Mourino, *Tetrahedron Lett.*, 1991, **32** 2813.
33. M.J. Dunn and R.F.W. Jackson, *J. Chem. Soc. Chem. Commun.*, 1992, 319.
34. L.A. Sarandeses, A. Mourino and J.-L. Luche, *J. Chem. Soc. Chem. Commun.* 1991, 818.
35. A P Marchand and G M Reddy, *Synthesis*, 1991, 198.
36. R. Sato, T. Nagaoka and M. Saito, *Tetrahedron Lett.*, 1990, **31** 4165.
37. B-H. Han and P. Boudjouk, *Organometallics*, 1983, **2** 769.
38. K. Imi, K. Imai and K. Utimoto, *Tetrahedron Lett.*, 1987, **28** 3127.
39. K.J. Moulton, S. Koritala and E. N. Frankel, *J. Amer. Oil Chem. Soc.*, 1983, **60** 1257.
40. A. Alexakis, N. Lensen and P. Mangeney, *Synlett,* 1991, **9** 625.
41. P. Boudjouk, B-H. Han, J.R. Jacobsen and B.J. Hauck, *J. Chem. Soc. Chem. Commun.*, 1991, 1424.
42. A. Tai, T. Kikukawa, T. Sugimura, Y. Inoue, T. Osawa and S. Fujii, *J. Chem. Soc. Chem. Commun.* 1991, 795.
43. R.S. Davidson, A. Safdar, J.D. Spencer and B. Robinson, *Ultrasonics*, 1987, **25** 35.
44. J. Ichihara and T. Hanafusa, *J. Chem. Soc. Chem. Commun.*, 1989, 1848.
45. J. Ichihara, K. Funabiki and T. Hanafusa, *Tetrahedron Lett.* 1990, **31** 3167.
46. J.C. Cochran and M.G. Melville, *Synth. Commun.*, 1990, **20** 609.
47. E.A. Schmittling and J.S. Sawyer, *Tetrahedron Lett.*, 1991, **32** 7207.
48. L.L. Adams and F.A. Luzzio, *J. Org. Chem.* 1989, **54** 5387.
49. M.J.S.M. Moreno, M.L. Sae Melo and A.S.C. Neves, *Tetrahedron Lett.*, 1991, **32** 3201.
50. T. Momose, N. Toyooka, H. Fujii and H. Yanagino, *Heterocycles,* 1989, **29** 453.
51. G. Olah and A-H. Wu, *Synthesis*, 1991, 204.
52. D. Goldsmith and J.J. Sorm, *Tetrahedron Lett.* 1991, **32** 2457.
53. J. Pan, I. Hanna and J-Y. Lallemand, *Tetrahedron Lett.* 1991, **32** 7543.
54. L-Y. Chen and L. Ghosez, *Tetrahedron Lett.* 1990, **31** 4467.
55. J.M. Khuruna, P.K. Sahoo and G. Maikap, *Synth. Commun.,* 1990, **20** 2267.
56. B.C. Ranu and M.K. Basu, *Tetrahedron Lett.* 1991, **32** 3243.

Ultrasound in Synthesis: Sonochemistry as a Tool for Organic Chemistry

Caroline M.R. Low
JAMES BLACK FOUNDATION, 68 HALF MOON LANE, LONDON SE24 9JE, UK

1 INTRODUCTION

The effects of ultrasound on chemical and biological systems have been under investigation for over 70 years and yet its use as a tool for the synthetic chemist scarcely predates 1980. A steady stream of papers has appeared since that point in time which clearly indicate that this is a technique with great potential for use in synthesis[1,2]. Its appeal lies in the speed with which complicated reactions can be carried out under extremely mild conditions without the need for esoteric apparatus. In fact the only apparatus required for the vast majority of reactions described in the literature is the standard ultrasonic cleaning bath that is to be found in most laboratories.

The aim of this paper is to investigate the potential that ultrasound represents as a tool for synthesis by illustrating some of the success that we have had in applying it to a range of different reaction systems. This technique has allowed us to extend the scope of certain reactions but, most importantly, has also led to the development of new chemistry that provides concise and efficient routes to complex molecules. The examples below are drawn from work carried out at the James Black Foundation and Professor Steven Ley's laboratories at Imperial College, London.

Synthesis of complex organic molecules typically involves the use of long reaction sequences. In particular, it is clearly important that selective transformations can be carried out on late stage intermediates under mild conditions. For instance, the reagents sodium phenylselenide and samarium diiodide have been used in a number of elegant syntheses and yet their preparation requires the use of high grade reagents and the rigorous exclusion of both air and water over long periods of time required to generate the active species. Conversely, the equivalent sonochemical reaction takes a fraction of the time and produces quantitative yields of reagent from off-the-shelf reagents without the need for inert atmospheres or pre-dried solvents. The beneficial effects of adding electron transfer agents to these systems became apparent during the course of these investigations and prompted us to examine recent proposals concerning the classification of sonochemical reactions on the basis of their mechanisms. These suggest that reactions whose rates, or product distributions, can be modified by ultrasonic irradiation correspond to those which proceed through radical intermediates[3-5]. This led us to investigate the effects of ultrasound on a typical cycloaddition reaction which would not be expected to fall into this category, in this case the Diels-Alder reaction between maleic anhydride and anthracene.

The second section concerns the role that ultrasound plays in promoting the reaction between diiron nonacarbonyl and a variety of intermediates to provide a route to π-allyltricarbonyliron lactone complexes. These ferrilactone complexes have been shown to be versatile intermediates in the synthesis of both 4-, 5-, and 6-membered lactones and lactams, and this methodology has been used to provide concise routes to both β-lactam and δ-lactone antibiotics. Furthermore, the δ-lactone products can be used as precursors of benzenesulphonyl ethers *via* the respective lactol. The benzenesulphonylethers form the basis of versatile methodology that has been established for the formation of bonds at this anomeric position. For example, this strategy has been used to prepare the spiroacetal functional group that is found in many biologically active compounds such as insect pheromones, ionophores, fungicides, insecticides and antiparasitic agents.

In particular, both of these strategies have been combined to provide a highly convergent synthesis of the anthelmintic macrolide Avermectin B1a. The molecule contains a 16-membered ring, 20 stereogenic centres and an array of chemically sensitive functional groups and gives a clear indication of the potential of ultrasound for use as a synthetic tool.

2 ACCELERATING SONOCHEMICAL REACTIONS WITH ELECTRON TRANSFER AGENTS

Sodium phenylselenide is a commonly used reagent for the introduction of seleno substituents into alkyl halides and sulphonates[6,7] However, it is not commercially available and must be prepared directly prior to use. Furthermore many of the published routes to this and related reagents involve the use of extremely odiferous reagents and toxic or incompatible solvent systems.

The direct reaction between diphenyldiselenide and sodium metal is heterogeneous and consequently the thermal reaction between these substrates is negligible at ambient temperature. There are no established criteria for identifying reaction systems that might benefit from exposure to ultrasound. Despite this, the weight of evidence from the existing body of literature clearly shows that ultrasound has profound effects on similar reactions, particularly those that involve metals such as magnesium, lithium, and zinc, and hence we decided to investigate the effects of ultrasound on this system.

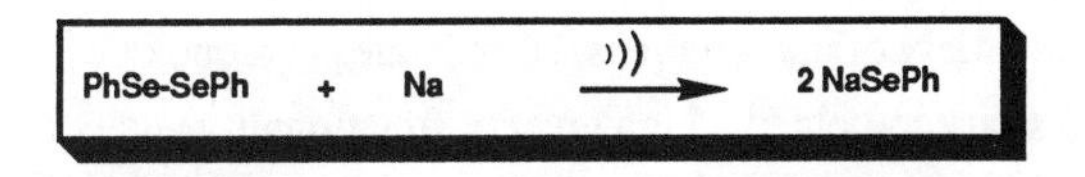

Reaction Conditions	Reaction Time/h
1. Solid Na, THF	72
2. Solid Na, xylene	31
3. Commercial Na dispersion, THF	21
4. Commercial Na dispersion, Ph_2CO, THF	0.1

Scheme 1

Sonicating a mixture of diphenyldiselenide and sodium in THF gave a quantitative yield of the reagent after 72h (Scheme 1)[8]. The reaction time was shown to be strongly dependent on the available surface area of the metal and could be reduced in one of two ways : firstly use of xylene in place of THF reduces aggregation of the dispersed metal[9] . However using xylene as a solvent would limit the method to the preparation of involatile and thermally stable phenylselenoethers. Further reductions in the reaction time were obtained by using a commercial 50% dispersion of sodium in paraffin[10] with the result that the reaction now took 21h to come to completion. This move also had the benecicial effect of allowing us to revert to using our preferred choice of solvent, THF.

At this point we had reduced the reaction time by a total of 70 % from its original 72 h by optimising the available surface area of the sodium metal, and further improvements required a different approach. Since the rate limiting step of this reaction is transfer of electrons from the metal surface to the diselenide in solution, we posed the following question : could the rate of the sonochemical reaction be increased by adding an electron transfer agent to the system? We chose to use benzophenone for this purpose to generate the deep blue sodium benzophenone ketyl *in situ.* This had a dramatic effect on the rate of the reaction and dropwise addition of a solution of diphenyldiselenide to the reaction mixture resulted in instantaneous formation of sodium phenylselenide. In addition the reaction became self-indicating as the deep blue suspension of sodium and benzophenone in THF gradually decolorises during the addition of the yellow diphenyldiselenide. The reaction was complete within five minutes once the last traces of mauve had disappeared from the cream-coloured suspension of sodium phenylselenide[8].

The reagent generated under these conditions produced consistent results when reacted with a variety of alkyl halides, sulphonates and epoxides - even in cases where the substrate contained a high level of functionality and competing reactions might have been expected to prove problematic. For example, a fragment of the ionophore antibiotic Tetronasin (1) was prepared using this methodology[11]. The highly functionalised tetrahydrofuran (3) was a key intermediate in the total synthesis of the natural product. However, all attempts to convert the diol (2) to the secondary alcohol (3) by direct reduction of a primary mesylate substituent had failed. Similarly, attempts to convert the mesylate to a phenyl selenoether using literature methods had also proven unsuccessful. However, reaction of the mesylate with sodium phenylselenide generated under our sonochemical conditions gave a quantitative yield of the selenoether, which was immediately reduced with Raney nickel to give the desired compound (3) as a single diastereoisomer (Scheme 2).

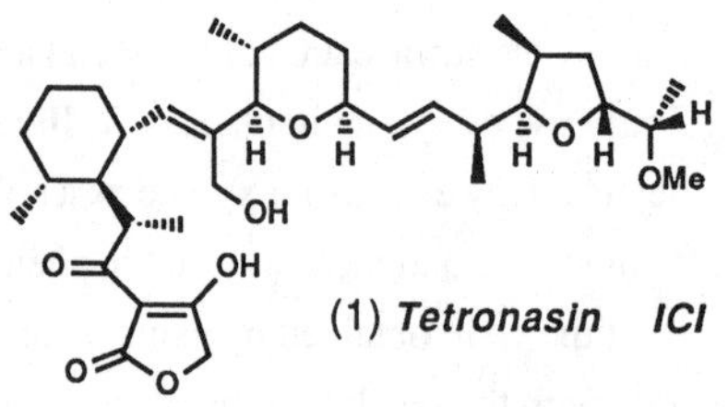

(1) ***Tetronasin ICI 139603***

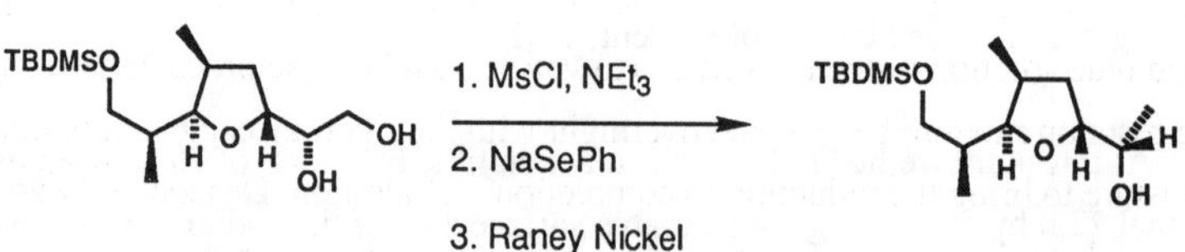

(2) → (3)

1. MsCl, NEt_3
2. NaSePh
3. Raney Nickel

Scheme 2

Having established methodology for the preparation of sodium phenylselenide we turned our attention to the reagent samarium diiodide (SmI_2). This reagent has generated a great deal of attention in the past 5 years as a result of the diverse nature of the reactions that it mediates[12]. It is a powerful reducing agent and initiates a variety of selective coupling reactions and functional group conversions with halogen- and oxygen-containing substrates. Organosamarium compounds have also been shown to exhibit similar reactivity to the widely-used alkyllithium and Grignard reagents. Kagan first described the preparation of SmI_2

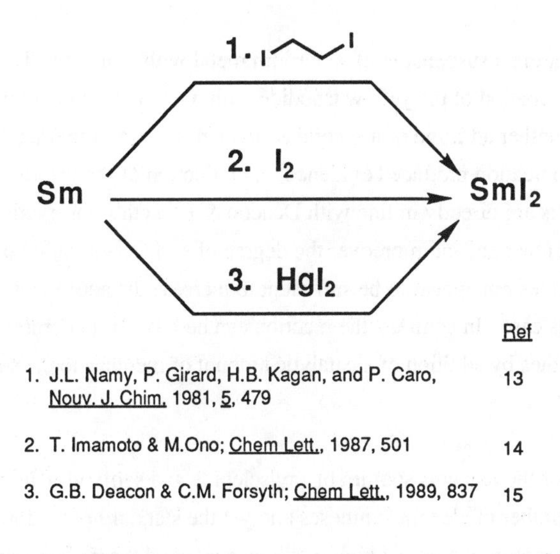

Scheme 3

from the reaction of samarium metal with 1,2-diiodoethane (Scheme 3) [13]. However, the procedure required that the reaction be carried out in a Schlenck tube with rigorous exclusion of air and water from the system - to the extent that the THF solvent had to be distilled twice under an inert atmosphere and the 1,2-diiodoethane carefully purified before use. These factors clearly limit the appeal of the reagent for use on a routine basis, and the direct reaction of samarium with iodine was later exploited [14,15] to provide access to the reagent as a 0.1M solution in THF. The first formed product in these last two cases is actually the yellow triiodide, $SmI_3(THF)_3$, that is reduced *in situ* by excess samarium [14] or precipitated mercury [15] to give a deep blue solution of the diiodide. However, both of these procedures require that the reaction mixture be refluxed overnight with the rigorous exclusion of both air and moisture to bring the reduction to completion. In addition, Deacon and Forsyth have shown that the results of using samarium metal as the reducing agent are strongly dependent on the quality of the metal used[15].

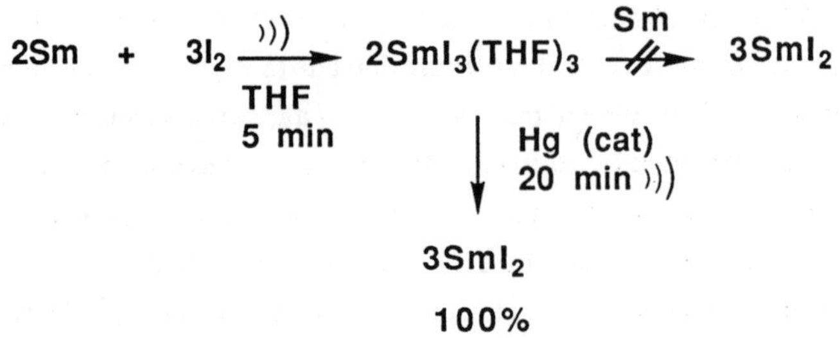

$$2Sm + 3I_2 \xrightarrow[\text{THF, 5 min}]{)))} 2SmI_3(THF)_3 \overset{Sm}{\not\longrightarrow} 3SmI_2$$

$$2SmI_3(THF)_3 \xrightarrow{\text{Hg (cat), 20 min)))}} 3SmI_2 \quad 100\%$$

Scheme 4

Sonicating a suspension of samarium metal with iodine in THF results in complete formation of the yellow triiodide within a five minute period (Scheme 4) However, neither addition of a second equivalent of samarium metal, nor a further 2h period of sonication produced evidence that reduction of the triiodide was occurring. These results are directly in line with Deacon & Forsyth's observations concerning the thermal reaction, and moreover the degree of surface activation produced by sonication does not appear to be sufficient to increase the activity of the samariun metal in this case. In contrast, the reaction can be brought to completion within twenty minutes by addition of a catalytic amount of mercury to the sonicated reaction mixture [16].

Both of the reagents sodium phenylselenide and samarium diiodide have been used in a number of elegant syntheses and yet the standard procedures for their preparation require the use of high grade reagents and the rigorous exclusion of both air and water over the long periods of time required to generate the active species. In each case the equivalent sonochemical reaction is complete within a fraction of the time and provides quantitative yields of reagent from off-the-shelf reagents without the need for exclusion of air, or use of pre-dried solvents. Whilst surface effects are clearly important in both of these heterogeneous reaction systems, these studies have also demonstrated that the rate of these sonochemical reactions can be dramatically increased by the presence of electron transfer agents - benzophenone in the first case and mercury in the second. These effects are much greater than would be anticipated on the basis of the equivalent thermal reactions and suggest that ultrasound has a special role to play in enhancing the rate of electron transfer reactions.

3 THE EFFECTS OF ULTRASOUND ON THE DIELS-ALDER REACTION

These observations concerning the effect of electron transfer agents on sonochemical reactions are in agreement with Prof. Luche's proposals[3-5] that "true" sonochemical reactions are those in which a single electron transfer is involved in the key step. However, the corollary of these proposals is that ionic and two-electron transfer processes should be essentially insensitive to cavitational phenomena. The majority of literature examples in which ultrasound has been shown to have profound effects involve reactions with metals, and it seems likely that such organometallic reactions involve single electron transfer (SET) processes. Nevertheless, there are also a small number of reports of ultrasound accelerating ionic processes, such as nucleophilic addition reactions[1], and cycloaddition reactions[17-21] which not be expected to fall into this category. For example, Lee and Snyder have reported a series of cases in which ultrasound has been shown to have a dramatic effect on the rate of a Diels-Alder reaction (Scheme 5)[18-21]. In each case the dienophile was the *o*-quinone (4) which was employed as the common intermediate in the synthesis of a series of abietanoid pigments . These compounds are natural products isolated from the Chinese sage, *Salvia miltiorrhiza Bunge*, and are of interest because they have been shown to be responsible for the biological activity of the Chinese traditional medicine Dan Shen. In each case, the yield of the standard thermal reaction could be substantially increased by carrying the reaction out under high pressure (entry 3), or sonicating the reaction mixture in a cleaning bath (entry 6). Both these procedures gave similar yields of the adducts (5) & (6) within a shorter period of time and also, somewhat fortuitously, altered the product distribution in favour of the desired regioisomer (5).

4 + → (5) + (6)

Reaction conditions	% Yield	Product ratio (5) : (6)
1. 110°C, benzene, 12h, sealed tube	<10	-- : --
2. reflux, MeOH, 16h	40	2.5 : 1
3. 11kBar, MeOH, 2h, RT	67	6 : 1
4. Eu(fod)$_3$, 0.08 eq, benzene	31	10 : 1
5. Eu(fod)$_3$, 0.08 eq, MeOH	20	10 : 1
6.))) , neat, 45°C, 2h	65	7 : 2

Scheme 5

However, it is generally accepted that the classical Diels-Alder reaction between a conjugated diene and an olefin is a concerted process that occurs in a single step via a six-centred transition state. As such, these results appear to run contrary to Luche's proposals that a two-electron process should be insensitive to cavitational effects. In fact, whilst it is generally accepted that the majority of the evidence points to the reaction occurring as a concerted process, two other mechanisms have also been considered. Both of these alternatives involves stepwise formation of the two new C-C bonds *via* either diradical or diion intermediates. In this particular case it could be suggested that the benzofuran-4,5-dione dieneophile (4) is capable of stabilising a diradical intermediate and that the role of ultrasound is to switch the reaction pathway away from that followed by the thermal reaction to an alternative SET process. It is clearly difficult to provide unequivocal evidence in favour of either proposal and so we chose to investigate a simpler system that would not be expected to diverge from the classical concerted reaction pathway.

The reaction of anthracene with maleic anhydride in toluene was examined under a variety of conditions[22] (Figure 1). The reaction mixture is homogeneous above 50°C and so the effects of the ultrasound are not obscured by any of the "mechanical" effects of cavitation. In this case use of a 55kHz cleaning bath as the ultrasound source proved ineffective and a low yield of bicyclooctane (7) was obtained over a 5h period. However, the reaction could be brought to completion within 4h when a direct immersion microtip probe was used. The internal temperature of the reaction was found to rise to 60°C during the course of sonication and this required that a control reaction be carried out at this temperature in the absence of ultrasound. The results show that the equivalent thermal reaction is considerably slower than the sonochemical reaction and was only 11% complete after the same period of time. Hence, ultrasound is producing a definite rate increase of the [$4\pi+2\pi$] cycloaddition reaction. At this point it should be noted that the thermal reaction in refluxing toluene at 110 °C is faster than the sonochemical reaction. Nevertheless, the mild nature of the sonochemical reaction conditions may well prove to be of interest from a synthetic point of view when considering reactions with substrates of limited thermal stability.

Conversely, the possibility that the course of the reaction had now switched from a concerted path to one in which a single electron transfer mechanism was now dominant could not be ruled out. We have clearly shown that addition of electron transfer agents increases the rate of a number of sonochemical reactions that would be

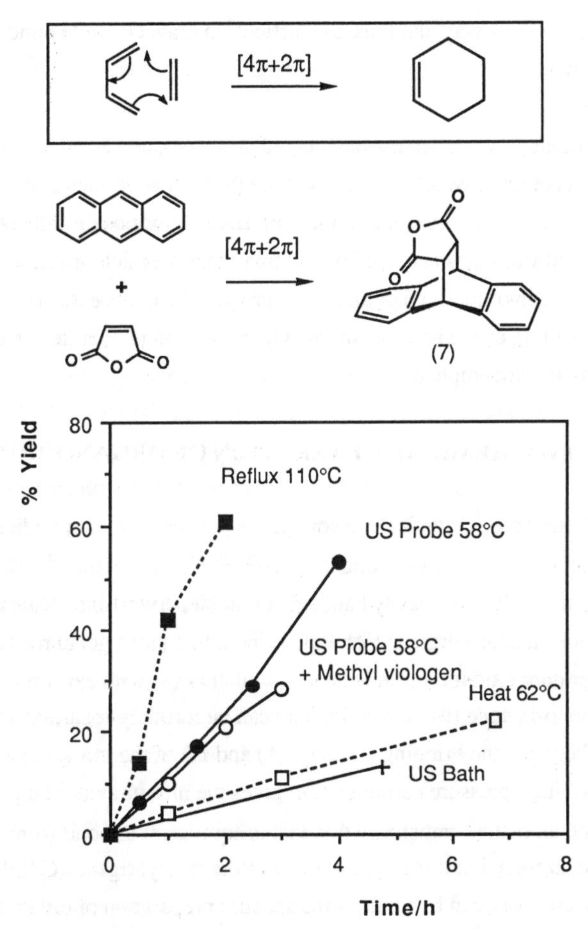

Figure 1 The Diels Alder reaction of anthracene and maleic anhydride

expected to involve single electron transfer. Hence, we chose to examine the effects of adding a similar agent, in this case Methyl Viologen, on the course of this reaction. The choice of Methyl Viologen was made for two reasons : firstly that it is known to participate in single electron transfer processes and has been shown to behave as both a redox indicator and an artificial electron carrier in enzyme reactions[23]; and secondly that the intermediate radical is bright red and the presence of such a highly coloured species would have been clearly visible to the naked eye. The results clearly show that addition of this electron transfer agent had no effect on the course of the sonochemical reaction and no change in the colour of the reaction mixture was observed. This seems to preclude the presence of radical species, although it should be stressed that the choice of Methyl Viologen for this purpose was not optimised in any way and thus it is difficult to draw concrete conclusions from this experiment.

Nevertheless, we have demonstrated that ultrasound has a clear effect on the rate of this reaction in which the mechanical effects of acoustic cavitation are not in operation. These results are not in line with Luche's proposals and could not have been predicted from his classification of "true" sonochemical reactions. Hence, rigorous adoption of these guidelines may prove to be unnecessarily restrictive whilst our understanding of the mechanism by which ultrasound operates on chemical systems remains incomplete.

4 SONOCHEMICAL GENERATION OF ORGANOIRON COMPLEXES

π-Allyltricarbonyliron lactone complexes are versatile intermediates that provide concise routes to a variety of cyclic lactones[24,25,26] and lactams[27]. The ferrilactone complexes themselves are easily handled, air-stable, crystalline solids that can be isolated using standard organic techniques, including silica gel chromatography. Low temperature oxidation of ferrilactones, such as (8),with cerium (IV) gives good yields of the β-lactone (9) - a reaction that can be formally regarded as bond formation between the bridging carbonyl (*) and C2 of the π-allyl ligand. Conversely, high pressure carbonylation gives the unsaturated δ-lactone (10) as the sole product, an overall transformation that is equivalent to bond formation between the bridging carbonyl and the opposite end of the π-allyl ligand (C4). Hence, a single ferrilactone complex can be used for the specific preparation of either β- or δ-lactones (Scheme 6).

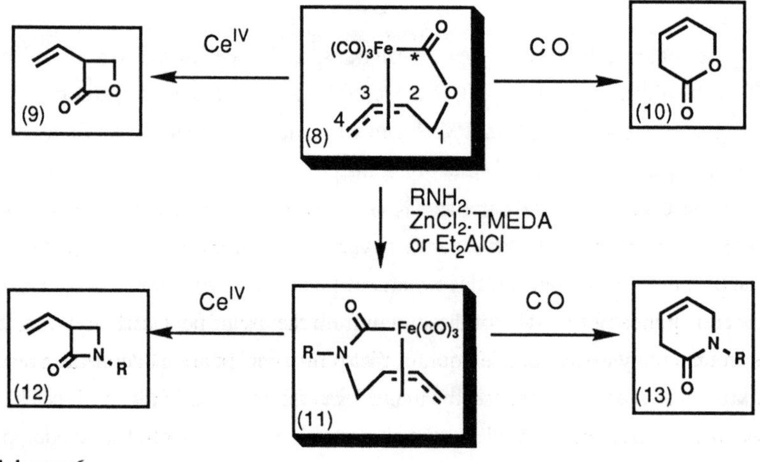

Scheme 6

Reaction of the π-allyltricarbonyliron lactone complexes with amines in the presence of a Lewis acid catalyst, such as $ZnCl_2$.TMEDA or diethylaluminium chloride, gives the equivalent ferrilactam complex (11) and treatment with cerium (IV), or high pressure carbonylation then gives access to the 4-ring (12), or 6-ring lactams (13) in a directly analogous way. Hence, these iron complexes are versatile intermediates that provide concise routes to a wide range of useful compounds.

The ferrilactone complexes themselves are generated from the reaction of coordinatively unsaturated tetracarbonyliron species with a variety of substrates which includes vinyl epoxides [24,26,28], Z-butene-1,4- diols[29,30] or the equivalent halogen compounds (Scheme 7)[31]. Alkenylcyclic sulphites have also been used as equivalents of epoxides and can be generated *in situ* from the reaction of the appropriate 1,2-diol with thionyl chloride[32]. and α,β-unsaturated sulphites such as (14) give good yields of ferrilactone under sonochemical conditions.[33]

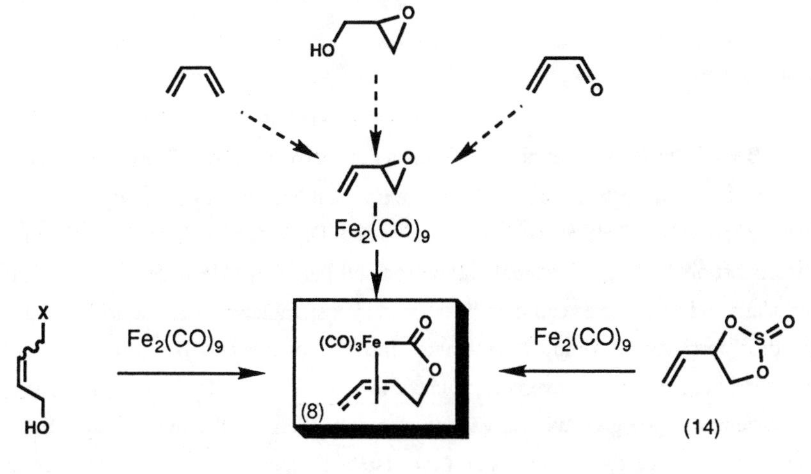

HO
O
O
O
$Fe_2(CO)_9$
X
$Fe_2(CO)_9$
$(CO)_3Fe$
O
O
(8)
$Fe_2(CO)_9$
O
S
O
O
HO
(14)

X = OH, Cl, Br

Scheme 7

The original literature conditions for generation of ferrilactone complexes involved the reaction of a vinyl epoxide with pentacarbonyl iron under photochemical conditions[26,29,31]. However, this procedure has several disadvantages that limit its general applicability. Most of these stem from the use of pentacarbonyl iron, which is an extremely toxic, volatile liquid . Yields of complex are moderate as a result of difficulties associated with cooling the reaction mixture and the thermal stability of the products. These problems prompted us to develop two alternative methodologies for the generation of $Fe(CO)_4$ (Scheme 8).

Preparation of ferrilactone complex (8)

Substrate	Reaction conditions	% Yield
vinyl epoxide	$Fe(CO)_5$, PhH, hν	62
	$Fe_2(CO)_9$, THF, 1h (reflux)	94
	$Fe_2(CO)_9$, PhH, 1h,)))	76
but-2-ene-1,4-diol	$Fe_2(CO)_9$, THF, 1h (reflux)	73
	$Fe_2(CO)_9$, PhH, 1h,)))	55
vinyl cyclic sulfite	$Fe_2(CO)_9$, PhH, 4-12h,)))	95

Scheme 8

Both of these involve the use of diiron nonacarbonyl ($Fe_2(CO)_9$), which is a solid and consequently a much easier reagent to handle. The diiron complex is partially soluble in THF at room temperature and IR studies suggest that the complex disproportionates to give pentacarbonyliron and $Fe(CO)_4$·THF in solution. [34] The tetracarbonyliron species can then be trapped by a suitable substrate to afford high yields of ferrilactone complex. However, this reaction does not occur in non-coordinating solvents. Conversely, sonicating slurries of $Fe_2(CO)_9$ and vinyl epoxide in benzene solution gives virtually identical results (Scheme 8). In this case the intermediate tetracarbonyliron species can also be trapped by 1,3-conjugated dienes to afford virtually quantitative yields of the η^4-tricarbonyliron(diene) complexes [35] (15) (Scheme 9). It is interesting to note that the yield obtained from the thermal reaction with $Fe(CO)_4$.THF is substantially reduced and is typically only 10-15%. The diene complexes are interesting synthetic intermediates in their own right as complexation to the tricarbonyliron moiety has the effect of reducing the reactivity of the diene system towards hydrogenation, electrophilic attack and Diels-Alder reaction. This feature has been exploited in its use as one of the few groups that can be used for protecting a conjugated diene system[36] The sonochemical synthesis of these complexes is much more efficient than existing literature methods which involve heating a mixture of the diene and $Fe(CO)_5$ in a high-boiling, inert solvent for long periods of time, with the result that yields of complex obtained rarely exceed 35%[37]. Conversely, the sonochemical reaction conditions are so mild that the product obtained from reaction with β–ionone was predominantly the exocyclic complex (16) after 1h. Continued sonolysis, or standing at room temperature led to isomerisation to the endocyclic complex (17), which is presumably the thermodynamic product of the reaction. In fact, this appears to be the only case in which isolation of such a kinetic product has been reported[35].

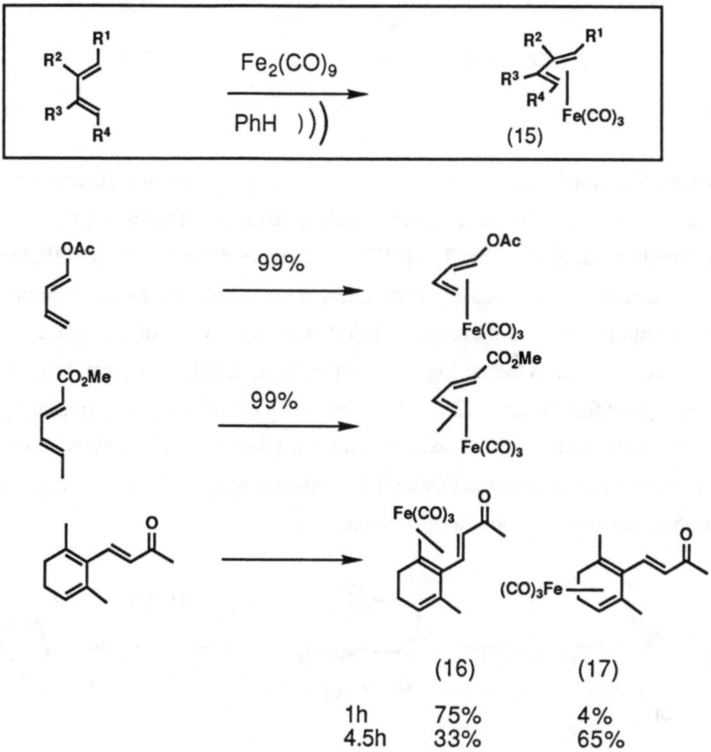

R¹ R² R³ R⁴ $Fe_2(CO)_9$ PhH))) → (15) $Fe(CO)_3$

OAc 99% → OAc $Fe(CO)_3$

CO_2Me 99% → CO_2Me $Fe(CO)_3$

→ $Fe(CO)_3$ O (16) $(CO)_3Fe$ O (17)

	(16)	(17)
1h	75%	4%
4.5h	33%	65%

Scheme 9

In general, the yields of ferrilactone complex obtained from the thermal reaction with $Fe(CO)_4$.THF and the sonochemical reaction with $Fe_2(CO)_9$ in benzene are broadly comparable and we have only found one case to date in which the paths of the two reactions diverged significantly (Scheme 10)[30]. The thermal reaction of the isobutene-1,3-diol with $Fe_2(CO)_9$ in THF solution gave the trimethylenemethane complex (18) as the major product and only a 12% yield of the expected ferrilactone (19) was obtained. Conversely, the product distribution from the sonochemical reaction was completely reversed such that the ferrilactone (19) now predominated and only trace amounts of the trimethylenemethane complex (18) were obtained. This difference was attributed to the influence of the

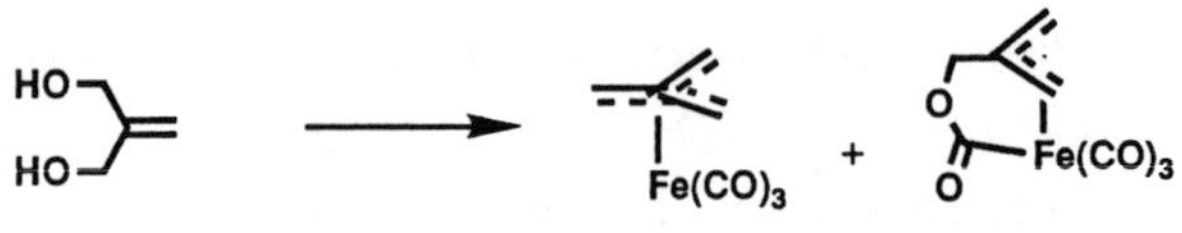

	(18)	(19)
$Fe_2(CO)_9$, THF, 1h	70	12
$Fe_2(CO)_9$, PhH, 1h,)))	trace	60

Scheme 10

reaction medium on the Lewis acidity of the tetracarbonyliron, which was reasoned to be greater in a non-coordinating solvent, such as benzene. This proposal was later substantiated by addition of $ZnCl_2$.TMEDA to the thermal reaction in THF, and the resultant increase in Lewis acidity completely reversed the product distribution. More importantly, ferrilactones such as (19) can be oxidised with cerium (IV) to afford 5-ring lactones (Scheme 11). In addition these ferrilactones can be converted to the analogous ferrilactams (20) using previously established methodology, such that the products of cerium (IV) oxidation are the γ-lactams. This extends the range of products that can be accessed using this methodology which now encompasses both 4-, 5- and 6-ring lactones and lactams.

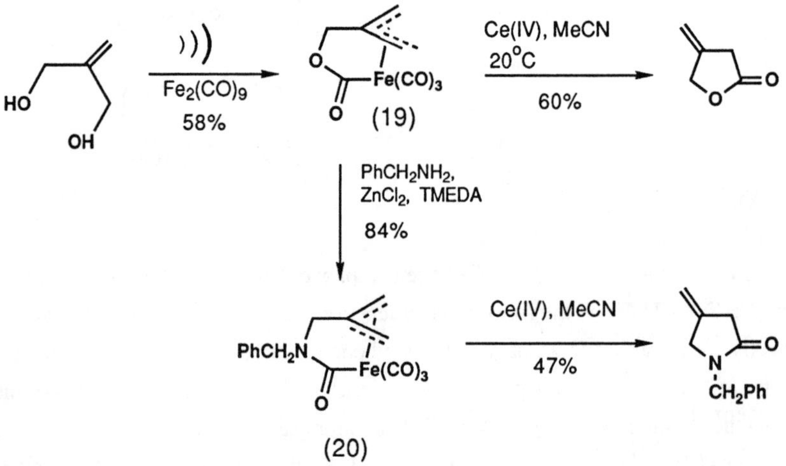

Scheme 11

This organoiron chemistry has now been applied to the synthesis of a range of pharmaceutically important compounds that include the β-lactam antibiotics, thienamycin[38] (21) and Nocardicin A[27] (22) (Scheme 12). However, it has proved most useful in allowing rapid entry to a whole series of δ-lactones, and this approach is illustrated in the synthesis of the

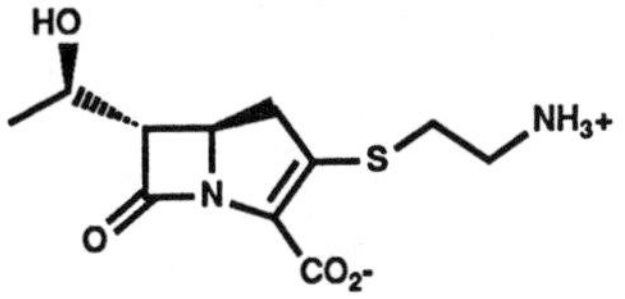

(21) (+)-Thienamycin 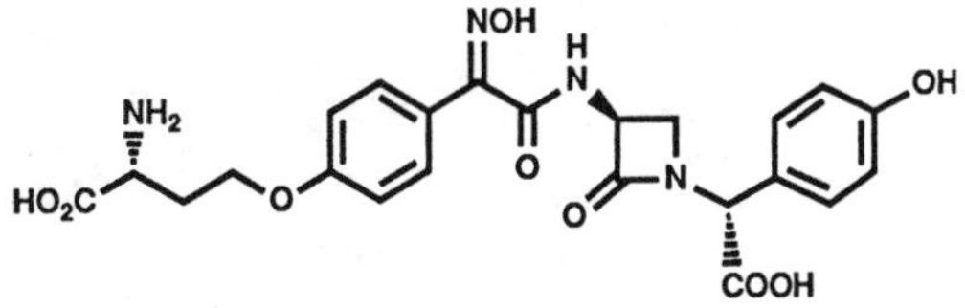 (22) Nocardicin A

Scheme 12

natural product Malyngolide (23). This compound was isolated from the blue-green algae *Lyngbya majascula* and has been shown to be active against *mycobacterium smegmatis* and *micrococcus pyogenes*[39,40] The vinyl epoxide (24) was prepared as illustrated in Scheme 13 and sonochemical reaction with $Fe_2(CO)_9$ in benzene gave good yields of the required ferrilactone complex (25). High pressure carbonylation gave the β,γ-unsaturated δ-lactone and the double bond was removed by hydrogenation in the presence of a PtO_2 catalyst. This process gave the natural product and its diastereoisomer as a 1:1 mixture that could be separated by column chromatography[25].

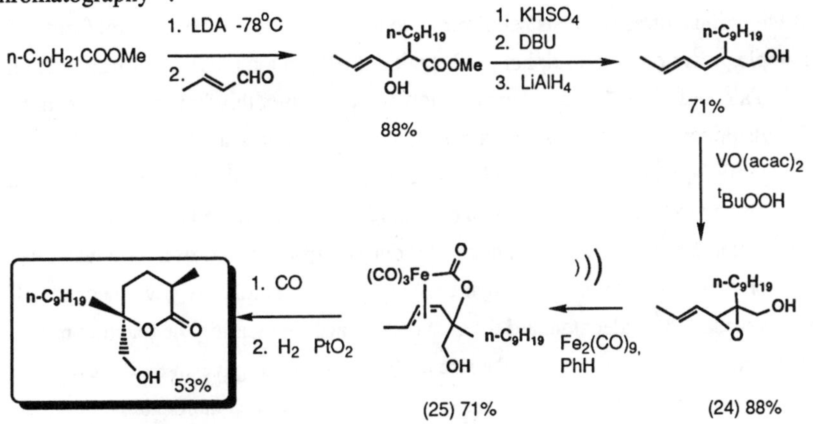

(23) Malyngolide

Scheme 13

The δ-lactone products of the sonochemical iron reaction can also be easily converted to crystalline 2-benzenesulphonyl ethers (26) by a simple sequence of steps that comprises hydrogenation of the double bond and reduction of the lactone carbonyl with DIBAL-H to give the unstable lactol, which is then trapped with phenylsulphenic acid (Scheme 14). These sulphones are versatile intermediates that can be used to introduce functionality at the anomeric centre of the tetrahydropyran unit, and their reactivity has been the subject of extensive studies within Prof. Ley's group. These results are summarised in Scheme 14. During the course of these

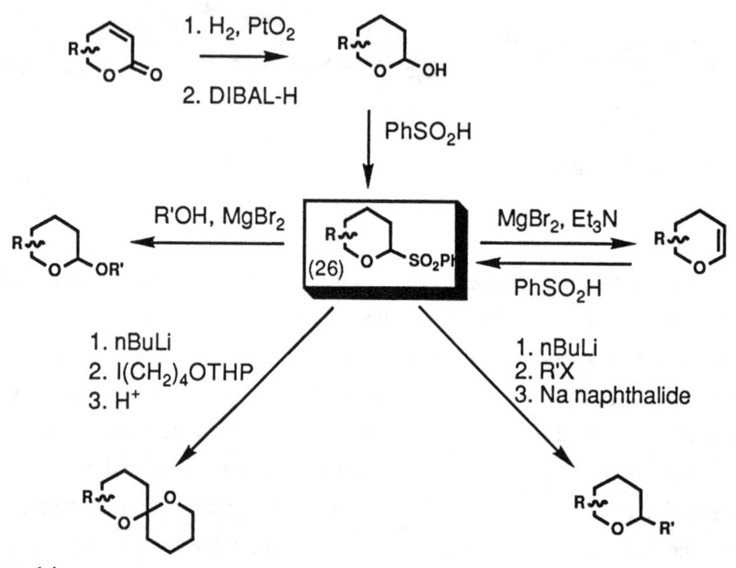

Scheme 14

investigations ultrasound has been shown to have a role to play on several fronts; for example, reaction of the sulphones (26) with $MgBr_2$ in the presence of triethylamine and an alcohol gives acetals. This reaction proceeds smoothly in the case of simple phenylsulphones and in most cases reaction was complete after 15-24 h at RT, or with gentle warming (50°C) in THF (Scheme 15). However, formation of glycosidic links proved to be more difficult as a consequence of the reduced reactivity of the highly functionalised phenylsulphone (27) derived from oleandrose. As a result the reaction between this protected sugar derivative and geraniol only gave modest yields of glycoside (28) under standard thermal conditions. Sonicating the reaction mixture produced a dramatic increase in the rate of the reaction and gave an 82% yield of the desired product. These results were subsequently found to be applicable to a number of cases in which the thermal reaction was sluggish and high yields of glycosides could be obtained under these mild conditions[41]and this reaction clearly shows potential for application in carbohydrate chemistry.

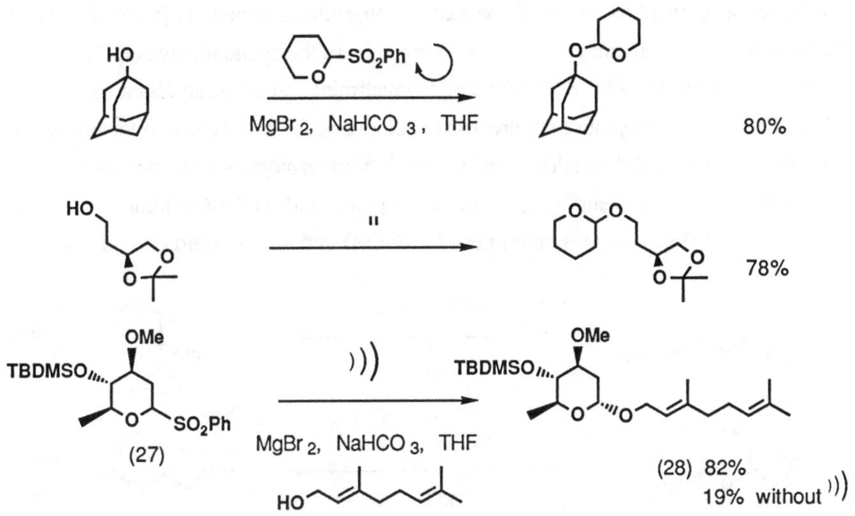

80%

78%

(27)

(28) 82%
19% without)))

Scheme 15

If we extend this to molecules in which the alcohol is an integral part of the system (29), the products of the intramolecular reaction in the presence of $MgBr_2.Et_2O$ will be cyclic ethers (30)(Scheme 16). Once again, the thermodynamically favoured 5- and 6-ring ethers can be formed without the intervention of ultrasound, but the 7- and 8-membered rings are only formed in the presence of ultrasound, and this represents a novel route to these medium-ring cyclic ethers[42].

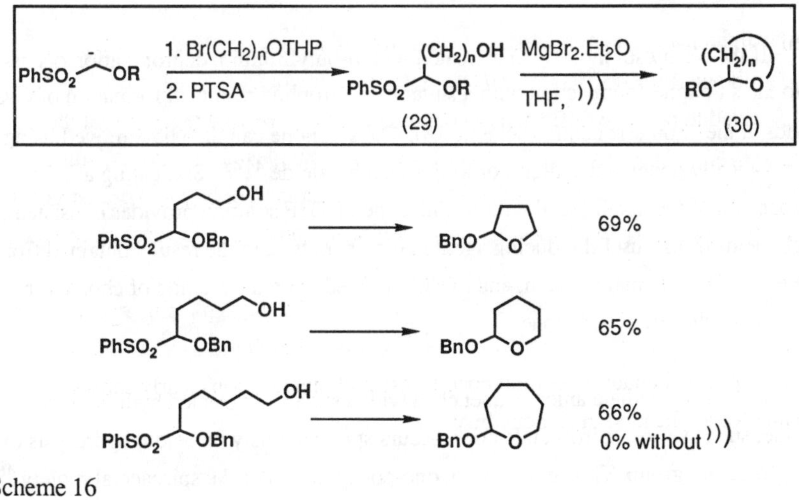

(29) (30)

69%

65%

66%
0% without)))

Scheme 16

If the sonochemical reaction with magnesium bromide is carried out in the absence of a suitable trapping agent, the products obtained from the cyclic phenylsulphonyl ethers are the glycals (31) & (32) arising from elimination of the sulphone (Scheme 17). This is not always an easy process to achieve with other substituents and use of ultrasound increased the yields significantly[43]. Furthermore, the reaction can be extended to the equivalent 2-phenylsulphonyl piperidines and pyrrolidines to provide a synthesis of the analogous enamines (33) & (34) under these mild conditions.

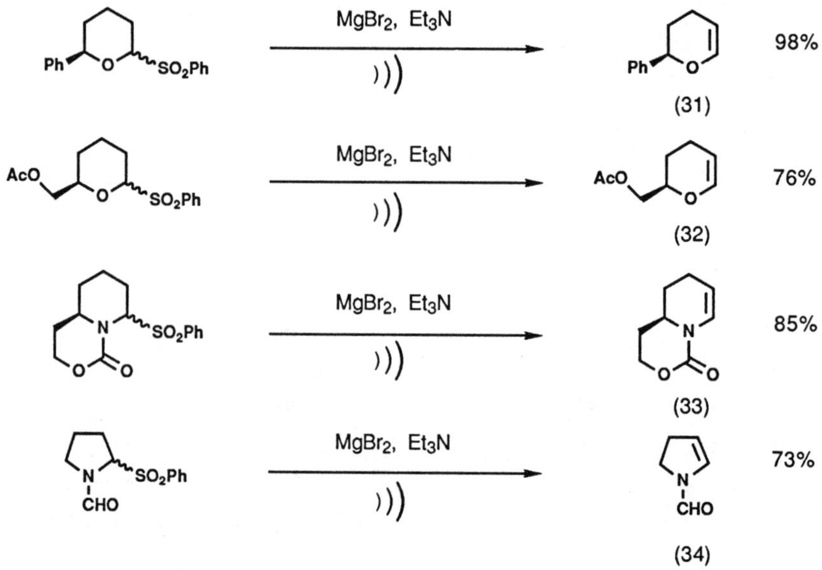

Ph, O, SO_2Ph — $MgBr_2$, Et_3N,))) → Ph, O (31) 98%

AcO, O, SO_2Ph — $MgBr_2$, Et_3N,))) → AcO, O (32) 76%

N, SO_2Ph, O, O — $MgBr_2$, Et_3N,))) → N, O, O (33) 85%

N, CHO, SO_2Ph — $MgBr_2$, Et_3N,))) → N, CHO (34) 73%

Scheme 17

If the phenylsulphone (26) is treated with n-butyllithium, deprotonation occurs α to the sulphone and reaction with a suitable electrophile results in formation of C-C bonds at the anomeric centre (Scheme 18). The sulphone can then be removed using ultrasonically generated sodium, or lithium naphthaleide [44,45]. Sonicating a suspension of the alkali metal with naphthalene in THF solution provides consistently high yields of this useful reducing agent in direct contrast to the results obtained from the equivalent thermal reaction, and this has proved to be the method of choice for removal of the phenylsulphone.

If the intermediate anion is quenched with a suitably protected hydroxyalkyl iodide, such as (36), spiroacetalisation occurs spontaneously on acidic hydrolysis of the protecting group. This represents a one-pot synthesis of the spiroacetal moiety [46,47] which is to be found in a wide range of ionophore antibiotics, insect pheromones and antiparasitic agents.

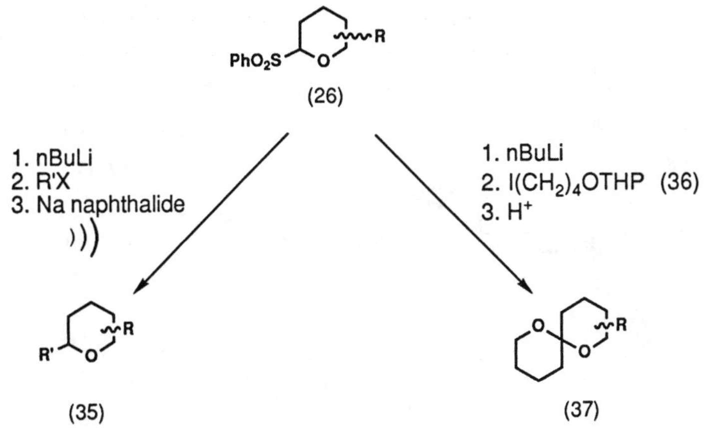

PhO_2S R

(26)

1. nBuLi
2. R'X
3. Na naphthalide
)))

1. nBuLi
2. $I(CH_2)_4OTHP$ (36)
3. H^+

R' R

(35)

R

(37)

Scheme 18

5 AVERMECTINS AND MILBEMYCINS

The previous section demonstrates the way in which ultrasound has been used to develop novel methodology on two distinct fronts . The first involved the generation of coordinatively-unsaturated transition metal carbonyl species which allowed us to develop chemistry associated with π-allyltricarbonyliron lactone complexes. This led to the development of novel syntheses of a range of 4-, 5-, and 6-membered lactones and lactams. Use of ultrasound has also allowed us to extend the reactivity of phenylsulphones and provide routes to glycosides, glycals and medium-ring cyclic ethers that are unavailable under thermal conditions. In each case application of ultrasound has allowed us to effect these transformations under extremely mild conditions and these reactions are applicable to a wide range of synthetic scenarios. In particular, both of these methodologies have been combined in the total synthesis of a family of natural products known as the Avermectins and Milbemycins.

These highly functionalised macrocycles are produced by a strain of Streptomyces and are potent anti-parasitic agents[48-51]. Each member of the family contains a spiroacetal functionality embedded in a 16-membered macrolide ring that contains an array of sensitive functional groups (Scheme 19). The potency and

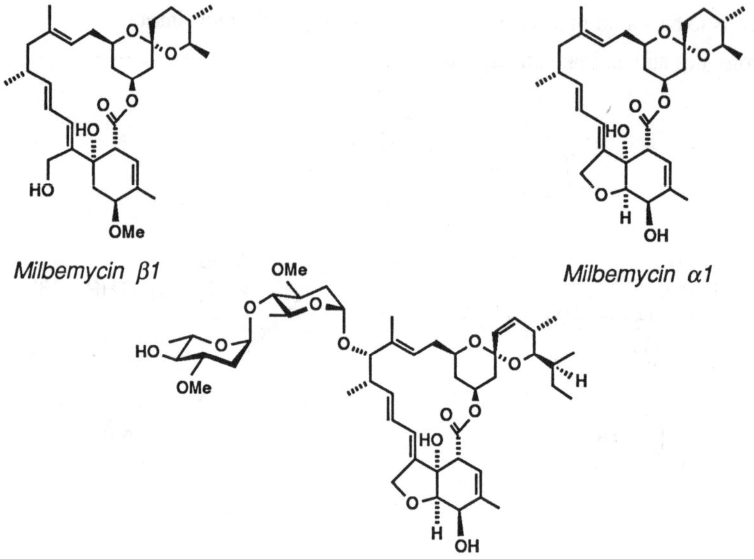

Milbemycin β1 *Milbemycin α1*

Avermectin B1a (38)

Scheme 19

specificity of these compounds is such that the 22,23-dihydroavermectin, Ivermectin, is marketed internationally as an anthelmintic despite its structural complexity, which is an order of magnitude greater than that found in most related agents. The macrocyclic aglycone of Avermectin B1A (38) itself comprises 20 stereogenic centres, 5 specifically-substituted olefinic units and an array of sensitive functional groups. In addition, the aglycone is coupled to a *bis*-oleandrose sugar moiety. As such, the synthesis of this molecule required a highly convergent strategy that would be equally applicable to the synthesis of the related milbemycins, whose biological activity is the subject of growing importance[50,51].

Close examination of the "Northern Hemisphere" of this compound suggests several openings for application of the chemistry described in the preceding sections of this review (Scheme 20). For example, both the *bis*-oleandrose and the spiroacetal moiety can be derived from a suitably functionalised δ-lactones and this presents an opportunity for the use of the sonochemical iron chemistry. The optically active ferrilactone complexes (40) & (41) were prepared from the sonochemical reaction of the cyclic sulphite (39) with $Fe_2(CO)_9$ in benzene in 65% yield[52]. The diastereoisomeric mixture of complexes was then carbonylated at 230 atm CO and 70°C in the presence of acrolein - which acts as a scavenger of the pentacarbonyliron

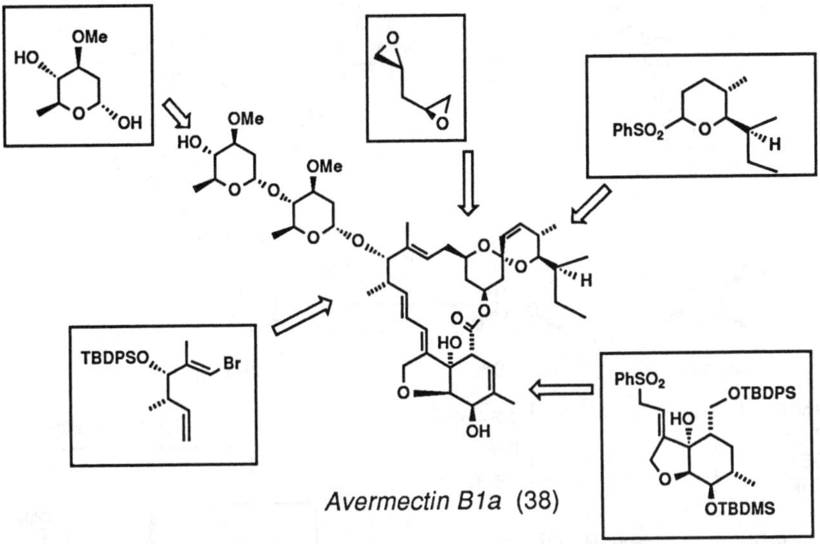

Avermectin B1a (38)

Scheme 20

that is formed during the course of the reaction. This proved to be essential to prevent isomerisation of the double bond into conjugation with the lactone carbonyl. The double bond of lactone (42) was then epoxidised using dimethyldioxirane and the unstable intermediates opened to give the allylic alcohols (43) & (44) in 26 and 49% yields. The alcohol (43) is a natural product called Osmundalactone, but for the purposes of this synthesis we needed to epimerise this centre under standard Mitsonobu reaction conditions. The lactones were then reduced to the unstable lactol (45) with DIBAL-H and treated directly with DBU in methanol at RT. Overnight equilibration gave oleandrose (46), and the opposite enantiomer cymarose (47) in 40 and 15% yields respectively. These products are clearly the result of ring-opening of the lactol, conjugate addition of methanol and reclosure to give a 70:30 equilibrium mixture of oleandrose and cymarose. This reaction could also be exploited to convert a further 60% of the unwanted cymarose to oleandrose in a separate experiment with the result that the combined yield of oleandrose from the lactones (43) & (44) was a respectable 40% following one recycle of the cymarose. Finally, the two oleandrose moieties were dimerised using a novel glycosidation procedure developed for this purpose which involves conversion of the C-4 monoacetate to the imidazolylcarbonyl glycoside and condensation with the analogous C-2 acetate in the presence of silver perchlorate.[52]

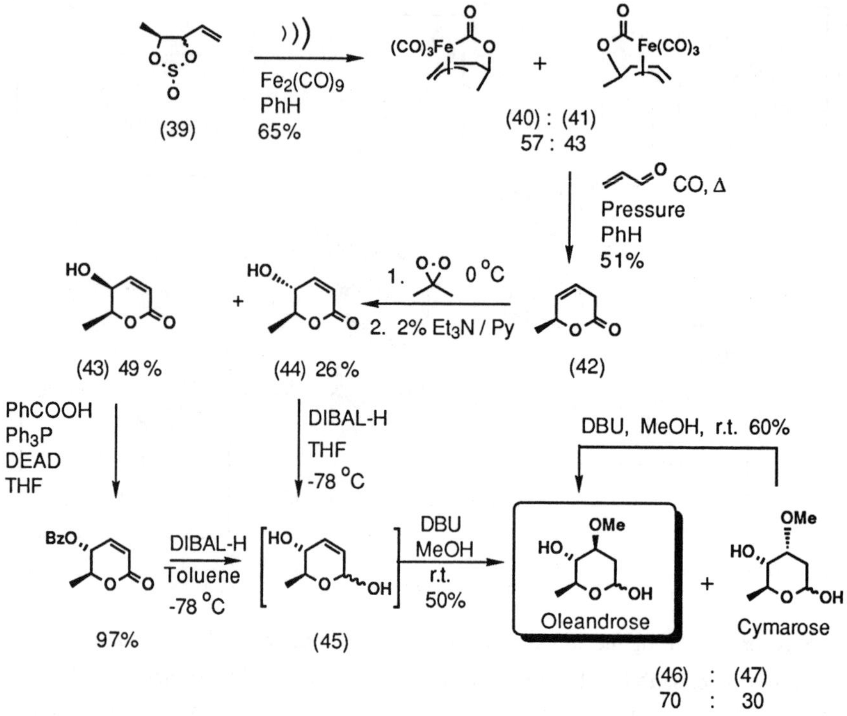

Scheme 21

The benzenesulphonyl ether (52) required for the synthesis of the spiroacetal unit was also prepared by carbonylation of a diastereoisomeric mixture of ferrilactone complexes (49) & (50), which were in turn prepared by the sonochemical reaction of $Fe_2(CO)_9$ with an optically active vinyl epoxide (48) (Scheme 22). The synthesis of the spiroacetal was then completed as illustrated in Scheme 23 : deprotonation with *n*BuLi at the anomeric centre and reaction of the anion with the epoxide (52) gave the enol ether (53). The TBDMS protecting group was then removed to allow spirocyclisation and the final sequence of steps introduced the required endocyclic double bond into the complete Northern Hemisphere of Avermectin B1a (54). This intermediate has now been elaborated to provide a total synthesis of the natural product (38).[52]

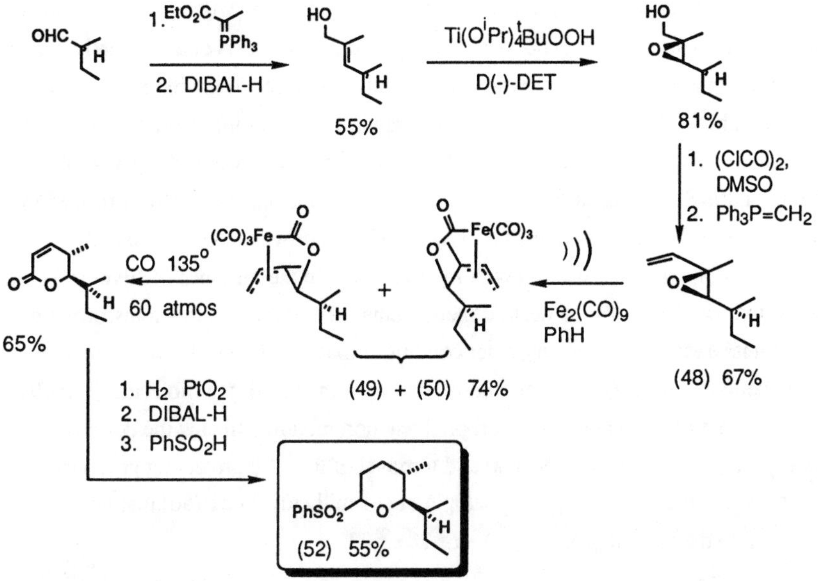

Scheme 22

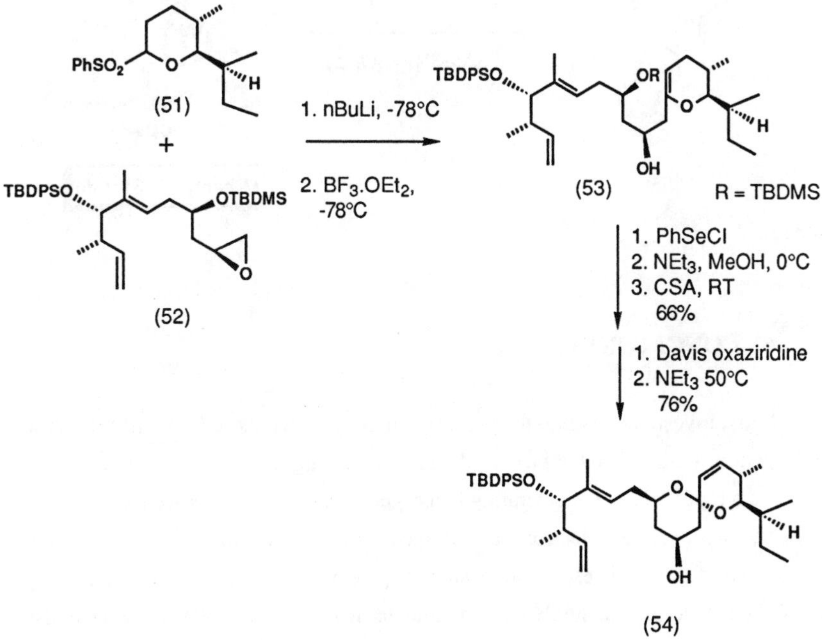

Scheme 23

An identical strategy was employed in the synthesis of Milbemycin β1 (56) and organoiron chemistry was successfully applied to the synthesis of an optically active benzenesulphonyl ether which was subsequently converted to the required spiroacetal.[53] In fact the synthesis was complete to the point where the only steps remaining were the methylation of the C-5 alcohol and removal of the silyl protecting group on the 8'-alcohol of the macrocycle (55). At this point we were surprised to find that what had appeared to be a trivial alkylation proved to be surprisingly difficult. For example, reaction with methyl iodide in the presence of silver (II) oxide failed to produce any of the desired methyl ether. Furthermore, this transformation could not be effected using any other conditions that would have been compatible with the sensitive array of functionality present in the rest of the molecule. Finally, we discovered that sonicating the original reaction mixture effected the desired methylation smoothly in 88% yield and removal of the silyl protecting group with HF/pyridine in acetonitrile gave a sample of (+)-milbemycin β1 (56) that was identical to the natural product (Scheme 24).[54]

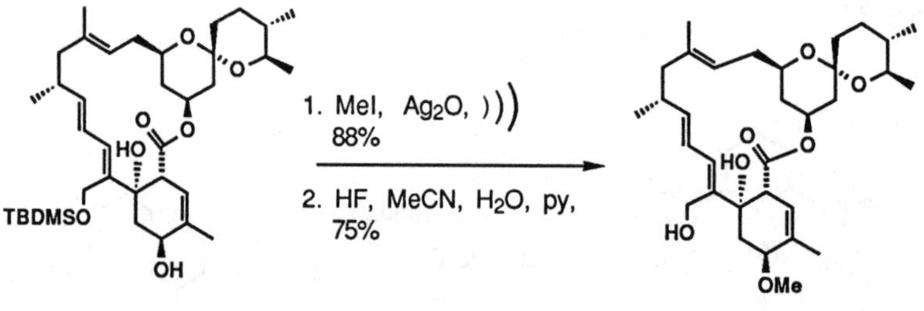

1. MeI, Ag_2O,)))
88%

2. HF, MeCN, H_2O, py,
75%

TBDMSO HO OH

(55)

HO HO OMe

Milbemycin β1 (56)

Scheme 24

6 CONCLUSIONS

Initial investigations into the preparation of the reagents sodium phenylselenide and samarium diiodide have produced some interesting results. Ultrasound has acquired a reputation for accelerating heterogeneous reactions, but we have shown that addition of suitable electron transfer agents increases the rate of the sonochemical reaction still further. These findings support proposals that ultrasound has a particular role to play in single electron transfer reactions. However, we have also shown that it has a positive effect on the rate of a simple Diels-Alder cycloaddition that would not have been expected to fall into this category. Hence, we must conclude that our understanding of the mechanism by which ultrasound acts remains incomplete. Nevertheless, this has not held back the development of new synthetic methodologies that exploit the beneficial effects of ultrasound on chemical systems.

Ultrasound has been applied to the reaction of various phenylsulphones with magnesium bromide to give a range of products that includes glycosides, glycals and medium-ring cyclic ethers. These are examples of reactions that do not occur at synthetically useful rates under thermal conditions, if at all. They have also used structurally complex substrates that would not have been expected to survive more forcing reaction conditions, and the total syntheses of the natural products Avermectin B1a and Milbemycin β1 provide further evidence that the sonochemical reaction conditions are fully compatible with highly functionalised molecules. Furthermore, they exploit novel organoiron chemistry which relies on the ability to generate the highly reactive tetracarbonyliron species under extremely mild conditions. The π-allyltricarbonyliron lactone complexes generated in this manner have been shown to be versatile intermediates in the synthesis of both 4-, 5-, and 6-membered lactones and lactams, and this methodology has been used to provide concise routes to both β- lactam antibiotics and the δ-lactone Malyngolide.

In conclusion, the aim of this paper has been to demonstrate that ultrasound is a tool worthy of serious attention from the synthetic community. This technique has allowed us to extend the scope of certain reactions but, most importantly, has led to the development of new chemistry that provides concise and efficient routes to complex molecules.

Acknowledgements

I would like to thank Prof. Steven Ley for all his help and encouragement and would also like to thank all the members of his group, past and present, who have contributed to this work.

REFERENCES

(1) S.V. Ley, and C.M.R. Low "Ultrasound in Synthesis"; Springer-Verlag: 1989; Vol. 27.

(2) C. Einhorn, J. Einhorn, and J.L. Luche, Synthesis 1989, 787.

(3) J.L. Luche, C. Einhorn, J. Einhorn, and J.V. Sinisterra-Gago, Tetrahedron Lett. 1990, 31, 4125.

(4) C. Einhorn, J. Einhorn, M.J. Dickens, and J.L. Luche, Tetrahedron Lett. 1990, 31, 4129.

(5) M.J. Dickens, and J.L. Luche, Tet. Letters 1991, 32, 4709.

(6) C. Paulmier "Selenium Reagents and Intermediates in Organic Synthesis"; Pergamon Press: 1986; Vol. 4.

(7) R. Monahan, D. Brown, L. Waykole, and D. Liotta In "Oraganoselenium Chemistry"; D. Liotta, Ed.; John Wiley & Sons: 1987; pp 207.

(8) S.V. Ley, I.A. O'Neil, and C.M.R. Low, Tetrahedron 1986, 42, 5363.

(9) J.L. Luche, C. Petrier, and C. Dupuy, Tetrahedron Lett. 1984, 25, 753.

(10) Aldrich Chemical Co. Ltd., Sodium 50 wt% dispersion in paraffin [7440-23-5].

(11) A.M. Doherty, and S.V.Ley, Tetrahedron Lett. 1986, 27, 105.

(12) J.A. Soderquist, Aldrichimica Acta 1991, 24, 15.

(13) J.L. Namy, P. Girard, H.B. Kagan, and P. Caro, Nouv. J. Chim. 1981, 5, 479.

(14) T. Imamoto, and M. Ono, Chem. Letts 1987, 501.

(15) G.B. Deacon, and C.M. Forsyth, Chem Letts 1989, 837.

(16) C.M.R. Low (Unpublished results 1992) Samarium powder (40 mesh, 99.9%) (0.9g, 6 mmol) was added to a solution of iodine (1.5g, 6mmol) in THF (10ml) and the mixture sonicated in a sealed flask (Bransonic Sonicleaner; 30W, 47kHz +/- 6%). After 5 mins formation of a yellow precipitate of the triiodide was complete and two drops of mercury were added. Sonication was continued for a further 20 min, after which the reaction mixture had turned deep blue and no trace of the triiodide remained.

(17) L.Y. Chen, and L. Ghosez, Tetrahedron Lett. 1990, 31, 4467.

(18) J. Lee, and J.K. Snyder, J. Org. Chem. 1990, 55, 4995.

(19) J. Lee, and J.K. Snyder, J. Am. Chem. Soc 1989, 111, 1522.

(20) J. Lee, H.S. Mei, and J.K. Snyder, J. Org. Chem. 1990, 55, 5013.

(21) M. Haiza, J.Lee, and J.K. Snyder, J. Org. Chem. 1990, 55, 5008.

(22) C.M.R. Low (Unpublished results 1992) All reactions were carried out using a 1:1 mixture of anthracene (10 mmol) and maleic anhydride (10 mmol) in toluene (50 ml). The course of the reaction was monitored by 300 MHz 1H NMR. The cleaning bath was a Bransonic Sonicleaner with a 30W output operating at 47 kHz +/- 6% and the Ultrasound Probe was a Lucas Dawe Ultrasonics Soniprobe 7534A fitted with a microtip direct immersion probe.

(23) For examples see J.V. O'Fallon, and R.W. Wright, Anal. Biochem 1991, 198, 179 and references therein.

(24) S.V. Ley, Phil. Trans. Roy. Soc. Lond. A 1988, 326, 663.

(25) A.M. Horton, and S.V. Ley, J. Organomet. Chem. 1985, 285, C17.

(26) R. Aumann, H. Ring, C. Kruger, and R. Goddard, Chem. Ber., 1979, 112,

(27) S.T. Hodgson, D.M. Hollinshead, S.V. Ley, C.M.R. Low, and D. J. Williams, J. Chem. Soc., Perkin Trans. 1 1985, 2375.

(28) K.-N. Chen, R.M. Moriarty, B.G. DeBoer, M.R. Churchill, and H.J.C. Yeh, J. Amer. Chem. Soc., 1975, 97, 5602.

(29) H.D. Murdoch, Helv. Chim. Acta. 1964, 47, 936.

(30) R.W. Bates, D. Diez-Martin, W.J. Kerr, J.G. Knight, S.V. Ley, and A. Sakellaridis, Tetrahedron 1990, 46, 4063.

(31) R.F. Heck, and C.R. Boss, J. Amer. Chem.Soc. 1964, 86, 2580.

(32) Y. Gao, and K.B. Sharpless, J. Amer. Chem. Soc 1988, 110, 7538.

(33) M. Caruso, J.G. Knight, and S.V. Ley, Synlett, 1990, 224.

(34) F.A. Cotton, and J.M. Troup, J.Amer.Chem.Soc. 1974, 96,, 3438.

(35) S.V. Ley, C.M.R. Low, and A.D. White, J. Organomet. Chem., 1986, 302, C13.

(36) T.W. Greene "Protective Groups in organic synthesis"; J. Wiley & Sons. Inc.: 1981.

(37) R. Pettit, and G.F. Emerson, Adv. Organomet. Chem. 1964, 1, 1.

(38) S.T. Hodgson, D.M. Hollinshead, and S.V. Ley, Tetrahedron 1985, 41, 5871.

(39) P.R. Burkholder, L.M. Burkholder, and L.M. Almodovar, Bot. Mar., 1960, 2, 149.

(40) T.J. Starr, E.F. Deig, K.K. Church, and M.B. Allen, Tex. Rep. Biol. Med. 1962, 20, 149.

(41) D.S. Brown, S.V. Ley, and S. Vile, Tetrahedron Lett. 1989, 29, 4873.

(42) P. Charreau, S.V. Ley, T.M. Vettiger, and S. Vile, Synlett, 1991, 415.

(43) D.S. Brown, P. Charreau, T. Hansson, and S.V. Ley, Tetrahedron 1991, 47, 1311.

(44) F. Jordan, P. Hemmes, S. Nishikawa, and M.Mashima, J. Am. Chem. Soc 1983, 105, 2055.

(45) T. Azuma, S. Yanagida, H. Sakurai, S. Sasa, and K. Yoshino, Synth. Commun. 1982, 12, 137.

(46) S.V. Ley, B. Lygo, F. Sternfeld, and A. Wonnacott, Tetrahedron 1986, 42, 4333.

(47) S.V. Ley, B. Lygo, and A. Wonnacott, Tetrahedron Lett. 1985, 26, 535.

(48) R.W. Burg, B.M. Miller, E.E. Baker, J. Birnbaum, S.A. Currie, R. Hartman, Y.-L. Kong, R.L. Monaghan, G. Olson, I. Putter, J.B. Tunac, H.

Wallick, E.O. Stapley, R. Oiwa, and S. Omura, Antimicrob. Agents. Chemother. 1979, 15, 361.
(49) T.W. Miller, L. Chaiet, D.J. Cole, L.J. Cole, J.E. Flor, R.T. Goegelman, V.P. Gullo, H. Joshua, A.J. Kempf, W.R. Krellwitz, R.L. Monaghan, K.E. Wilson, G. Albers-Schonberg, and I. Putter, Antimicrob. Agents. Chemother., 1979, 15, 368.
(50) H.G. Davies, and R.H. Green, Nat. Prod. Rep. 1986, 3, 87.
(51) M.H. Fisher, and H. Mrozik In "Macrolide Antibiotics,"; S. Omura, Ed.; Academic Press: Orlando, 1985; pp 533.
(52) S.V. Ley, A. Armstrong, D. Diez-Martin, M.J. Ford, P. Grice, J.G. Knight, H.C. Kolb, A. Madin, C.A. Marby, S. Mukherjee, A.N. Shaw, A.M.Z. Slawin, D.J. Williams, and M. Woods, J. Chem.Soc Perkin Trans I 1991, 667.
(53) C. Greck, P. Grice, S.V. Ley, and A. Wonnacott, Tetrahedron Lett. 1986, 27, 5277.
(54) N.J. Anthony, A. Armstrong, S.V. Ley, and A. Madin, Tetrahedron Lett. 1989, 30, 3209.

Ultrasonically Assisted Polymer Synthesis

Gareth J. Price
SCHOOL OF CHEMISTRY, UNIVERSITY OF BATH, CLAVERTON DOWN, BATH BA2 7AY, UK

1 INTRODUCTION

As is evident from the other Chapters in this book, ultrasound has been successfully applied to a wide range of synthetic procedures. It is therefore perhaps somewhat surprising that there is relatively little current effort in polymer sonochemistry, particularly given the commercial importance of macromolecular materials. In fact, the application of ultrasound to polymers predates other chemical applications. As long ago as the 1920's and 1930's, the reduction in the viscosity of solutions of natural polymers such as agar, starch and gelatin was noted by Szalay and Gyorgi[1-3]. Since then, ultrasound has been used in a number of areas of polymer chemistry.

High frequency ultrasound - in the range of 1 - 10 MHz - has been applied to the determination of structure and conformation of polymers as recently reviewed by Pethrick[4]. However, this chapter will deal only with those aspects that are appropriate for the preparation of polymeric materials, either by synthesis from low molecular weight monomers or by modification of pre-existing polymers. Polymer sonochemistry is a particularly rich area since, as will be seen, the opportunity exists to utilise all of the various effects caused by cavitation. We will attempt to review the subject to indicate the major areas of work at present with examples taken from work in the Author's laboratory where appropriate, and also to suggest areas where future developments are likely.

2 EXPERIMENTAL

The ultrasound techniques that we have employed in our syntheses are essentially identical to those used in other synthetic procedures. Some reactions were carried out using a reaction flask immersed in a Ney 2.8 L ultrasound bath operating at a nominal frequency of 35 kHz and power output of 110 W. Cold water was passed through a copper cooling coil in the bath to minimise any temperature rise. To obtain higher ultrasound intensities, reactions were performed with a "horn" system, either a 'Sonics and Materials VC50' or a MSE Soniprep being used. The reactants were contained in a cell designed and constructed in our laboratory and shown in Figure 1. It consists of a pear shaped flask modified by the addition of an indentation at the bottom to give good mixing and a water jacket for temperature control. Facility was also made for sampling via a gas tight syringe and for reaction under an inert atmosphere.

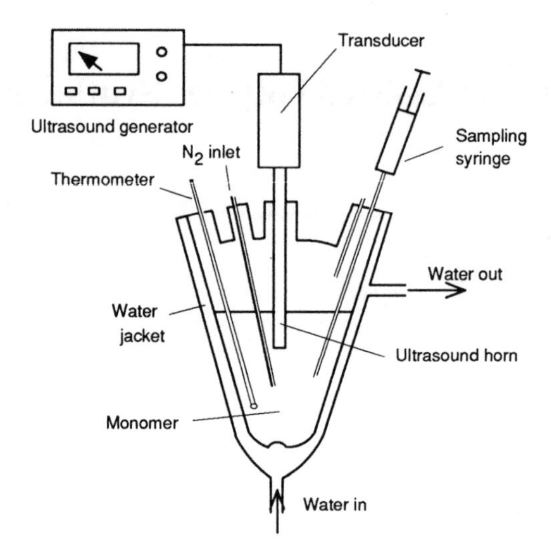

Figure 1. Cell for ultrasonic polymerization.

Clearly, in common with all chemists, we are interested in the rate and yield of our reactions and in the stereochemistry of our products. However, to fully characterise a polymer sample, we must also measure other properties such as its molecular weight and, particularly, the range and distribution of the molecular weights in the material. The latter property has important consequences for the physical and material properties of the polymer and is characterised by the polydispersity, γ, which takes a value of 1.0 if all the polymer chains are the same length (synthetically impossible at present) and increases with the breadth of the distribution. The usual way of measuring γ, as well as the molecular weights, is via Gel Permeation Chromatography[5]. This is a form of liquid chromatography which utilises a porous solid to separate the polymer chains according to their size to produce a chromatogram as shown in Figure 2 which is essentially a plot of concentration versus chain size. Since the size of the chain directly depends on its molecular weight, comparison of the chromatogram with those from calibration standards allows calculation of the various molecular weight averages as well as the breadth of the distribution. The GPC chromatography described here was performed with a Bruker LC21/41 gel permeation chromatograph using tetrahydrofuran as the eluent.

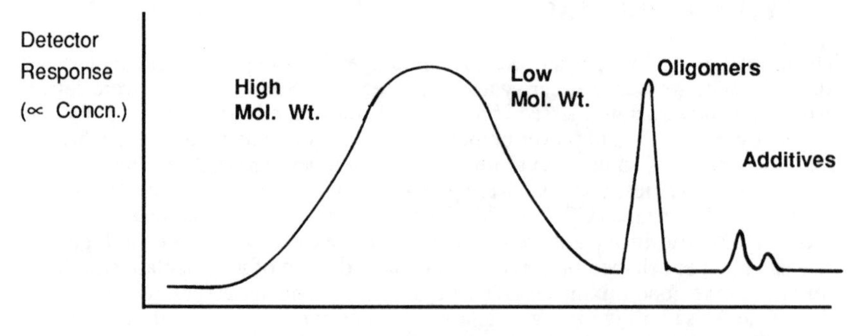

Figure 2. Schematic Gel Permeation Chromatogram.

3 SYNTHESES EMPLOYING ULTRASONIC DEGRADATION

As noted above, the reduction in molecular weight of polymer chains when exposed to ultrasound is perhaps the oldest of all known sonochemical effects. Although the nature of polymeric materials was not recognised at that time, it was later realised that the effect was due to breakage of the polymer chains and consequent degradation to lower molecular weights. The early reports prompted a large body of work over the succeeding two decades, reviewed in References 6 and 7, aimed at characterising the process in terms of the rate of bond cleavage of a wide range of polymers and the effect of the solution and ultrasound parameters. Further work has been carried out more recently and the process is now understood with sufficient detail to make it commercially viable in a number of areas. While there is still some debate about the precise origins of the degradation, the mechanism can briefly be described as the polymer chain being caught in the rapid flow of solvent molecules about collapsing and exploded cavitation bubbles and being subjected to extremely large shear forces. It might be thought that the extreme conditions of temperature found in cavitation bubbles would contribute to the degradation but, even after very long sonication times for polystyrene dissolved in a number of solvents, no styrene (the major product of thermal degradation) was detected[8]. In addition, thermal degradation produces cleavage in a random process while ultrasonic degradation is much more specific in that it has been shown by a number of workers that cleavage occurs preferentially near the middle of the chain[9,10] which is also consistent with the shear mechanism[11,12]. The basic effects of the process are shown in Figure 3 and indicate that the degradation is faster at higher molecular weights and reaches a limiting value, M_{lim}, below which no further degradation takes place. The rate and extent of degradation can be controlled by varying the solvent, the solution concentration, the ultrasound intensity as well as several other parameters[6,7,13-15] as demonstrated in Figure 4 so that we have a great deal of control over the process.

The aim of this chapter is to describe polymer synthesis and so it may come as a surprise to the reader to be told about a degradation process. However, exploitation of the control over this process allows the modification of existing polymers into new

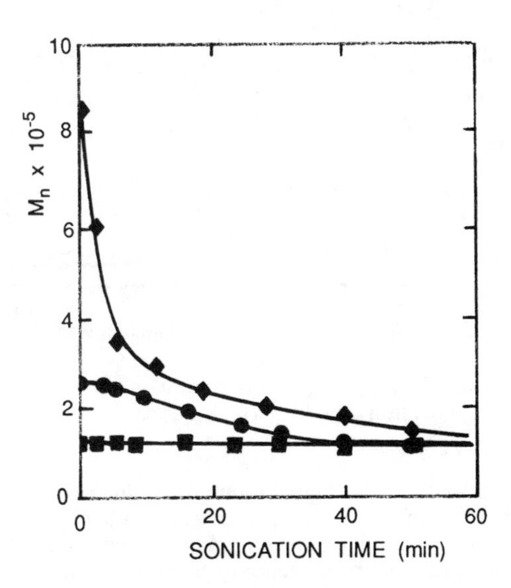

Figure 3. Ultrasonic degradation of 1.0 % (w/v) polystyrene solutions in toluene at 25 °C for three initial molecular weights.

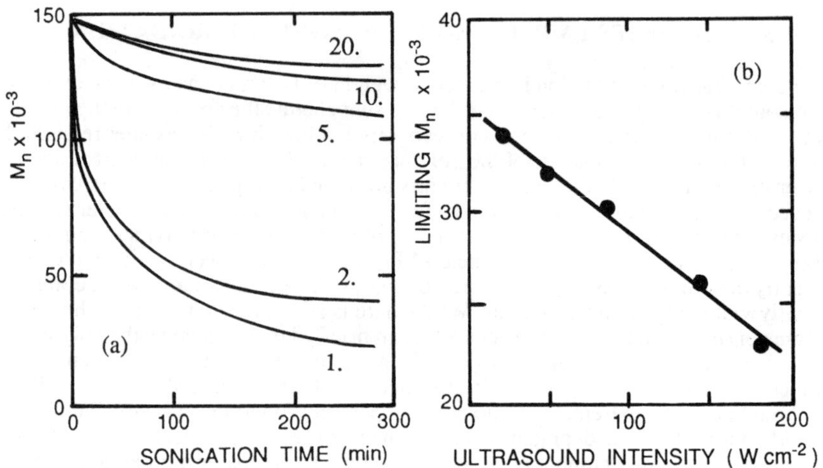

Figure 4. Ultrasonic degradation of polystyrene in toluene at 25 °C: (a). Effect of solution concentration (% w/v); (b). Effect of intensity.

materials. At its most straightforward level, the degradation can be used as an additional processing parameter to allow variation of the molecular weight distribution. For example, Figure 5 shows GPC chromatograms of a solution of Natural Rubber in toluene before and after sonication[16]. The degradation of the higher molecular weight species narrows the distribution markedly and this can have large effects on, for example, the viscosity and hence affect the further processing of the material. The use of higher ultrasonic intensities would have given greater changes in the same time.

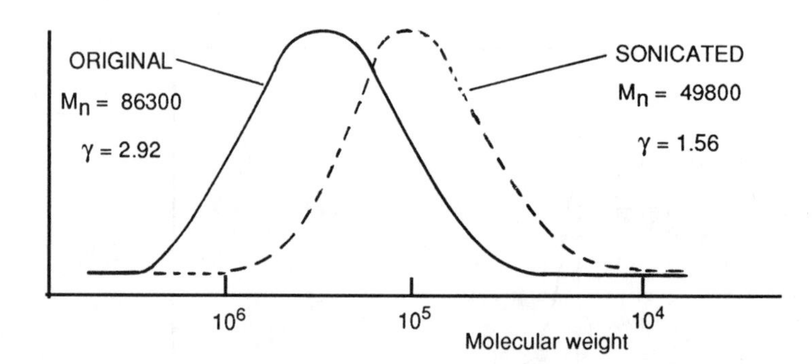

Figure 5. GPC chromatograms of natural rubber before and after 2 hr sonication as a 3% (w/v) solution in toluene.

A second application of the degradation utilises the macromolecular radicals formed as a consequence of the chain breakage. A number of workers have sonicated mixtures of two polymers dissolved in a common solvent, cross reaction between the two types of radicals forming a block copolymer, chains containing sequences of two (or more) different monomers which are used in a range of applications[17]. Incorporation of a second monomer can modify the solubility or elasticity of a polymer and block copolymers are finding large use as compatibilizers between different types of polymer. This approach was used by Melville and Murray[18] to show the presence of radical intermediates in the degradation. Henglein, as long ago as 1954[19,20] prepared poly(acrylonitrile-*b*-acrylamide) in aqueous solution and a large number of similar reactions have since been carried out[6,7] but they do suffer from difficulty in controlling and recovering the products. The approach which we have adopted is to sonicate a polymer dissolved in a solution containing the second monomer, demonstrated in Figure 6 for polystyrene and methyl methacrylate. From our degradation studies, we can control the structure of the first polymer quite precisely and also, by varying the concentration of monomer in solution, the block size of the second polymer, allowing a large degree of control over the material structure. An alternative approach is to sonicate the polymer in the presence of a species susceptible to radical attack where "end capped" polymers are formed. We have used this to prepare, for example, polystyrenes and poly(alkenes) bearing fluorescent groups[8].

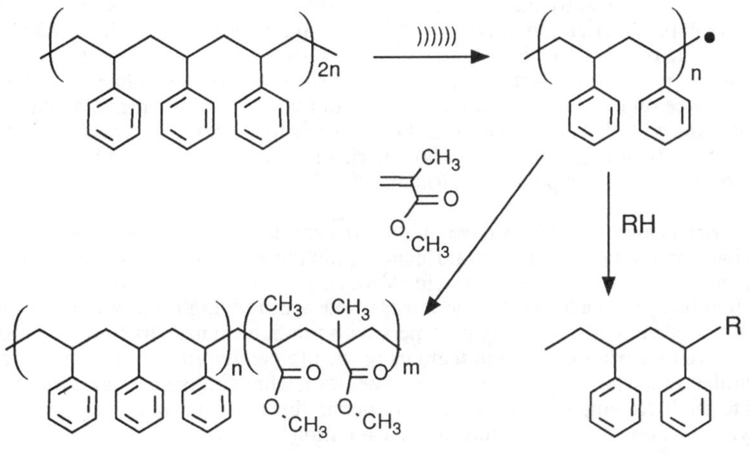

Figure 6. Ultrasonic production of end functionalised polymers and block copolymers.

4 RADICALLY INITIATED POLYMERIZATION

Despite considerable research into new techniques,[21] the most common method for the polymerization of vinyl monomers remains that using radical initiation. The initiating radicals are usually produced by thermal or photochemical decomposition either of the pure monomer or of an added initiator such as azo-bis *iso*butyronitrile, AIBN, or an organic peroxide[22]. Redox systems may also be used for aqueous reactions. A major focus of current work in this area is the control over the structure and properties of the resulting polymers. Although the molecular weights of polymers produced by radical initiation can be controlled, for instance by the addition of chain transfer agents, one of the major problems, caused by the high and indiscriminate reactivity of the growing radical, is the stereochemical control of the resulting polymer structure.

As noted in Chapter 1, sonication produces high concentrations of H• and OH• radicals in water[23-25] and the possibility of using ultrasound to initiate polymerization was suggested some time ago when Lindstrom and Lamm produced poly(acrylonitrile) in aqueous solution using this method[26], a system also studied by Henglein[19]. It is also now clear that radicals can be produced in organic liquids[27]. Clearly, if the ultrasonic method could be applied to vinyl monomers, it would provide an alternative method of initiation with the possibility of a great deal of control over the process.

Cavitation is much less efficient in organic liquids and it was long held that polymerization would not take place in organic systems. Indeed, El'Piner[28] stated that "*...polymerization of monomers in an ultrasonic field does not occur if these are thoroughly dried and do not contain substances in the polymerized state.*" This misconception, that no reaction would take place unless "seeded" with preformed polymer which would degrade to yield the initiating radicals continued for some time[29,30]. Despite this, some success was achieved by Melville who polymerized styrene, methyl methacrylate and vinyl acetate both in the presence and absence of poly(methyl methacrylate). Miyata and Nakashio[31] investigated the AIBN initiated polymerization of styrene and found that, while higher molecular weight polymers were formed, the polymerization rate decreased linearly with the ultrasonic intensity. However, subsequent studies by a number of workers cast doubt on this finding. Lorimer *et al.*[32] also found that there was an increase in molecular weight during the ultrasonic polymerization of N-vinyl carbazole but that there was an optimum intensity for the rate of reaction. The same workers[33] also investigated the polymerization of N-vinyl pyrrollidone a system that is unusual since it does not follow the usual kinetic dependence on monomer concentration due to formation of hydrogen bonded complexes with the solvent (water). On the basis that ultrasound was known to disrupt this type of bonding, an increased polymerization rate was expected but, experimentally, the opposite was found.

Kruus and co-workers have published a series of papers dealing with the mechanism and detailed factors influencing the ultrasonic bulk polymerization[34-38] of styrene (St) and methyl methacrylate (MMA), and work in the Author's laboratory[39-41] has involved these and similar monomers. We are studying ultrasonic initiation from the point of view of producing polymers with pre-determined structure and this work will serve to illustrate the main features of the ultrasonic process. Kruus also noted that a number of systems gave chars with some of the characteristics of a coal and attributed this to the high temperatures causing pyrolysis during transient cavitation, with normal polymerization occurring during stable cavitation[38].

The general features of the process are demonstrated in Figure 7. Figure 7(a) shows the variation of the number average molecular weight during the sonication of pure MMA at 25 °C at an ultrasonic intensity of 15.4 ± 0.5 W cm^{-2}. High molecular weight polymer is formed at early stages of the reaction but the average value at longer times falls exponentially. This is not the same plot as found with conventionally initiated radical polymerization[22] and is a consequence of the ultrasonic degradation process described above coming into operation on polymer chains once they have formed. The rate of polymer production and final yield under the same conditions are shown in Figure 7(b). A conversion of ~ 12% was achieved after 6 hr, similar to that achieved by Kruus *et al.* It was noticed that cavitation in the solution essentially stopped, shown by a marked change in the sound of the sonication at this time and no further conversion to polymer occurred thereafter. We ascribe this to the increased viscosity of the solution restricting movement of the solvent molecules and suppressing cavitation, hence preventing formation of radicals. It also demonstrates that cavitation is necessary for the production of polymer.

There are a large number of factors that influence the polymerization, the major ones being those such as the temperature and nature of the solvent which influence the cavitation in the liquid and also the intensity of the ultrasound which determines the number of cavitation events. Our aim is to completely characterize and develop a quantitative model of the polymerization so that we commenced with an in depth study of the ultrasonic initiation.

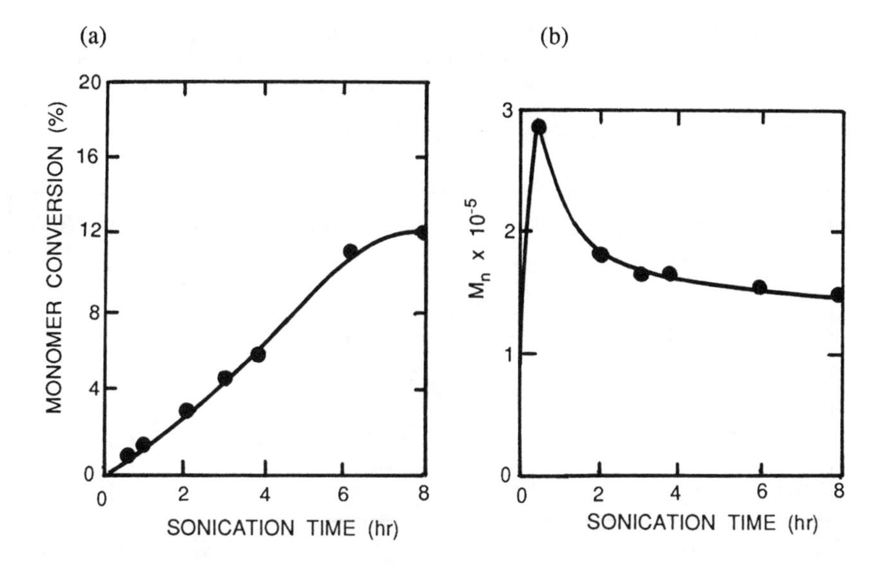

Figure 7. Ultrasonic polymerization of methyl methacrylate at 25 °C: (a). Conversion to polymer; (b). Variation in molecular weight.

The initiation process. To assist with determining the mechanism and in modelling the initiation, we needed a solvent that would behave in an ultrasound field in an identical manner to MMA except that it would not polymerise. As previously noted, the cavitation behaviour of a solvent is determined chiefly by its physical properties[42,43] such as viscosity, surface tension and, often considered to be the most important, vapour pressure. The liquid most closely matching MMA is methyl butyrate and the lack of a vinylic double bond prevents polymerization, confirmed by performing blank experiments. Thus, the application of ultrasound to these two "cavitationally similar" liquids would be expected to produce similar behaviour with dissolved polymers and also similar numbers of radicals on sonication.

The rates of initiation were estimated by trapping the radicals formed as a result of cavitation in methyl butyrate using an excess concentration of 2,2-diphenyl-1- picryl hydrazyl, DPPH. The use of a model solvent avoided complications from reactions with MMA and growing or degrading polymer chains. Obviously, there will be a degree of radical recombination and other side reactions, but since the radical must escape the cavitation bubble and solvent "primary cage" whether to initiate polymerization or to be trapped by DPPH, we feel that the rate of trapping will closely mimic the rate of initiation. By this method we were able to measure the rate constant for radical production under a range of conditions and the results are shown in Table 1.

Few studies of the thermal polymerization of MMA have been made but the rate of initiation has been estimated[44] to be 7×10^{-16} s^{-1} at 100 °C which is obviously negligible in comparison to our ultrasonic process. The ultrasonic rates are also much higher than the corresponding value for the thermal decomposition of AIBN in MMA at 25°C, obtained by extrapolating data from higher temperatures[45], is $\sim 2 \times 10^{-8}$ s^{-1}. Indeed, the Literature value for AIBN in MMA at 70°C is 3.1×10^{-5} s^{-1} so that, simply by using ultrasound alone, we can achieve at 25 °C similar rates of initiation as are usually found in thermal polymerizations. To further compare with more conventional experiments, we also sonicated a solution of 0.1 wt% AIBN in MeOBu, a concentration typical of that usually employed for polymerizations, the rate constant also being shown in Table 1. This is some three orders of magnitude higher than that expected for AIBN at this temperature so that the sonication process is clearly able to accelerate the decomposition of AIBN in solution as well as producing radicals directly.

Table 1. Rate constants for DPPH trapping at 25 °C

SOLVENT	TEMPERATURE (°C)	RATE CONSTANT 10^5 s^{-1}
Methyl butyrate	-10	6.35
Methyl butyrate	25	2.21
Methyl butyrate	60	1.03
Methyl butyrate + 0.1 % AIBN	25	9.13

A major aim of our work is to be able to control the rates of reaction so that we need to understand the effects of the various experimental parameters on the radical trapping. Figure 8(a) shows an Arrhenius plot for radical trapping in methyl butyrate, leading to an apparent activation energy of -19 kJ mol^{-1}. The negative value indicates that the process is faster at low temperatures and is not unusual in sonochemical reactions[42], being a consequence of the solvent properties being more favourable to cavitation at lower temperatures as explained below. It would be expected that the main parameter influencing the number of cavitation bubbles and hence the number of radicals produced would be the ultrasonic intensity, I_{us}[46] and its effect is shown in Figure 8(b). These indicate the expected linear dependence of the rate of DPPH

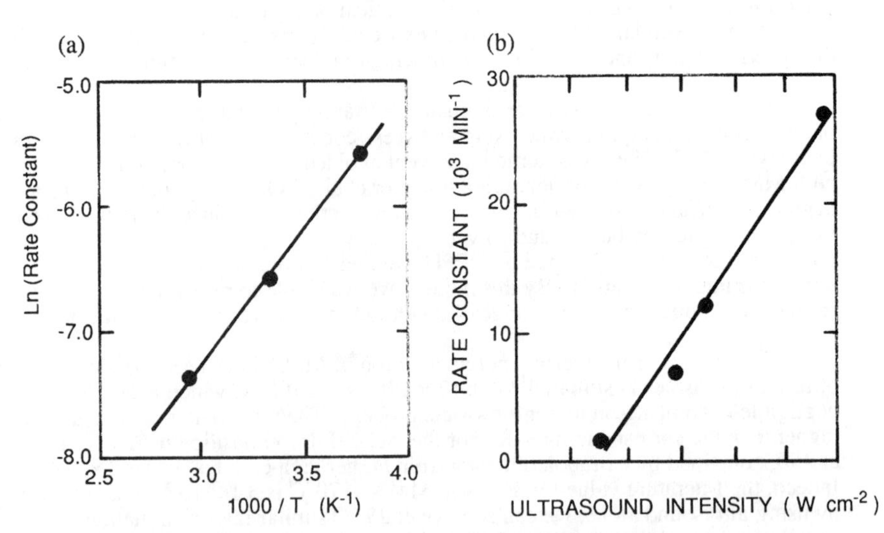

Figure 8. Sonochemical radical production in methyl butyrate at 25 °C: (a). Arrhenius plot; (b). Effect of ultrasonic intensity.

consumption on I_{us} and also show that there is a minimum intensity of approximately 12 - 13 W cm^{-2} below which no production of radicals takes place, presumably corresponding to the cavitation threshold in this system. Thus, by suitable manipulation of the conditions, we can control the rate of initiation.

The polymerization process. To develop a model of the polymerization, it is necessary to understand the effect of each of the factors described above on the process, including the propagation and termination reactions as well as initiation since these will determine the structure of the resulting polymer. In addition, while the conversion shown in Figure 7(a) is adequate to study the mechanism of the process, it is too low to be considered viable in a commercial sense so that we also considered various ways to increase the yield of polymer.

Polymerizations were carried out at a series of monomer concentrations in solution as well as over a range of temperatures. In each case, as in the work of Kruus *et al.*, we have been able to understand the effect of the various parameters and have been able to model the reaction kinetics using the usual type of steady state analysis[22,35], albeit with extra terms to account for the degradation process induced by ultrasound. As examples, the effect of temperature and ultrasound intensity on the rate of polymerization are shown in Figure 9. From all our results, we were able to conclude that the ultrasound has little, if any, significant effect on the propagation or termination reactions of growing radicals but only on the rate of radical production (assuming a factor of two or three in the diffusion rates to be insignificant compared to the several orders of magnitude acceleration of the initiation reactions).

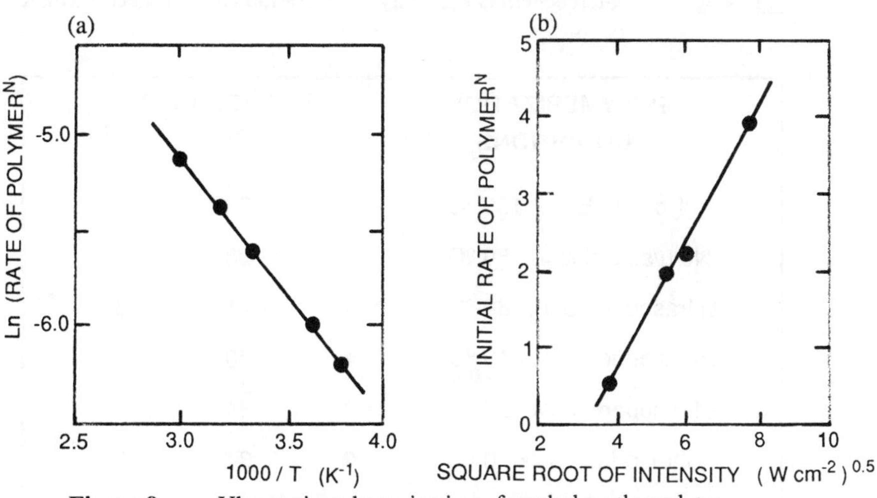

Figure 9. Ultrasonic polymerization of methyl methacrylate: (a). Arrhenius plot; (b). Effect of ultrasonic intensity at 25 °C.

To further characterize the ultrasonic process, we then investigated the microstructure of these polymers using NMR spectroscopy[47]. The stereochemistry of polymer chains is usually described by the *tacticity* which is assigned according to the arrangement of substituent groups along the chain as shown in Figure 10. *A*tactic polymers have a random arrangement while *iso*tactic and *syndio*tactic materials have the substituents all on the same side of the chain or in an alternating fashion. Bovey[47,48] has assigned the resonances due to the α-methyl groups in PMMA and the results from polymers produced in our ultrasound work are shown along with Literature results in Table 2.

As can be seen, conventional radical initiation using a peroxide initiator leads to predominantly syndiotactic polymers although there are significant sequences of atactic and isotactic groups. Our ultrasound promoted reactions at higher temperatures produce polymers with similar stereochemistry confirming that sonication has little or

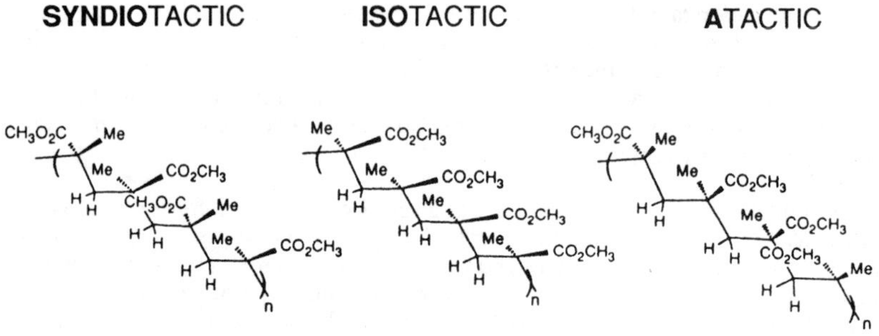

SYNDIOTACTIC **ISO**TACTIC **A**TACTIC

Figure 10. Sterochemical arrangement of substituents in PMMA.

Table 2. Stereochemical tacticity ratios for radically initiated PMMA

POLYMERIZATION CONDITIONS	TACTICITY (%) Iso-	A-	Syndio-
No u/s, Bulk, 100 °C	9	38	54
No u/s, Solution, 50 °C	6	38	56
Ultrasound, bulk, 60 °C	4	41	55
Ultrasound, bulk, 40 °C	4	40	56
Ultrasound, bulk, 25 °C	3	34	63
Ultrasound, bulk, 0 °C	2	34	65
Ultrasound, bulk, -10 °C	1	26	74

no effect on the propagation reaction. However, lowering the temperature raises the proportion of syndiotacticity along the chain. This can be explained since syndiotactic addition is thermodynamically slightly more favourable due to steric hindrance between the bulky ester groups. As the temperature is lowered, the propagation rate is slowed and there is more chance of the thermodynamically favoured addition taking place. Thus, ultrasound offers a chance to control the microstructure of the polymers.

Finally, following the increased rate of initiation under ultrasound illustrated above, reactions were carried out in the presence of an added initiator. A polymerization was carried out as above using a solution of 0.1 wt% AIBN in MMA. As expected from the foregoing discussion, the rate of polymerization was faster than in the absence of AIBN with a conversion of ~ 13 % being achieved in 4 hr. However, as in the reactions described above, polymerization essentially ceased at this point, again indicating that the process is viscosity limited.

5 EMULSION POLYMERIZATION

One of the most important commercial processes for preparing polymers is emulsion and/or suspension polymerization[49]. Here, small droplets of monomer are dispersed in an aqueous environment with the aid of surfactants and emulsifiers, and an initiator (either oil or water soluble) added to begin the polymerization. Thus, we have a very complex, multi-component system. The first to apply the well known emulsification properties of ultrasound to this type of reaction was Ostroski in 1950 who found increased dispersion of the various reagents. Hatate *et al.*[50,51] also studied this process using styrene as the monomer and found that ultrasound prevented the particles sticking to the walls of the container and also their agglomeration so minimising the build up of heat in the reactor. At the intensities used, they noted little effect on the kinetics of the polymerization. More recently, Lorimer and Mason's *et al.* group have also compared styrene polymerizations carried out in the presence and absence of ultrasound from a high intensity horn[52]. They found that the decomposition of potassium persulphate, an aqueous initiator, was accelerated and that this in turn, together with the production of more stable emulsions, increased the rate of polymerization. A lower molecular weight was produced than in the conventional reaction.

An alternative approach has been taken by Davidson and co-workers[53-55]. They employed an ultrasonic "whistle reactor" to prepare the emulsion while utilising a U.V. lamp to photoinitiate polymerization. Using this system with both oil and water soluble initiators gave latexes which were stable for several months even when conducted in the absence of surfactant. A smaller and more even size distribution was also obtained leading to systems, in their case adhesives, with superior properties to those obtained from a conventionally prepared emulsion.

6 RING OPENING POLYMERIZATIONS

A number of commercially important polymers are produced by a ring opening mechanism of a cyclic monomer[56]. Probably the most commercially significant in terms of amount produced is the reaction of ϵ-caprolactam to Nylon-6. In similar reactions, a range of polyesters can be produced from cyclic lactones[56].

The ϵ-caprolactam reaction has been studied by Ragaini *et al*[57,58]. For Nylon-6 production, the process is run in two stages, the initial ring opening catalysed by a small amount (~ 1%) of water, followed by polymerization under vacuum. Ragaini's work has showed that ultrasound allows a single step polymerization without the need to add water to start the reaction. In addition, no added water was needed so that commercially available ϵ-caprolactam could be used directly. High molecular weight materials were formed in shorter reaction times and a narrower distribution of molecular weights was found than when using the conventional process.

We have used as a model system the ring opening reaction of octamethyltetrasiloxane, catalysed by sulphuric acid, to poly(dimethyl siloxane), PDMS[59]. This is the base material of the large number of silicone products. Initially, we carried out 24 hour polymerizations at room temperature for an identical set of conditions[60] except that one was stirred and the other sonicated using a cleaning bath. The experiments gave similar yields: 71 % conversion to polymer for the stirred reaction and 68 % for the sonicated. However, the molecular weight distributions show considerable differences, as shown by the GPC traces in Figure 11. The reduction in polydispersity was an expected consequence of the ultrasonic degradation described above. However, this usually leads to a *reduction* in molecular weight so that our result was somewhat unexpected. Clearly, in the PDMS reaction the ultrasound is accelerating the ring opening reaction as well as affecting the molecular weight distribution.

$$[\mathrm{Me_2SiO}]_4 \xrightarrow[\text{)))) or stir}]{\mathrm{H^+},\ 30^\circ\mathrm{C}} \left[\mathrm{Si(Me)_2{-}O} \right]_n$$

In view of this preliminary result, we were interested to see whether sonication would allow us to reduce, or preferably, eliminate the acid catalyst. A series of polymerizations was carried out at various levels of catalyst and compared with unsonicated reactions under the same conditions. However, it was clear that the amount of catalyst was the main factor in determining the extent of reaction and, in particular, no polymerization occurred in the absence of added acid. To determine whether ultrasound of a higher intensity could initiate the ring opening in the absence of catalyst, we sonicated a pure sample of the cyclic tetramer on a horn system but found that a negligible amount of polymer was formed after a 24 hour period of insonation at room temperature.

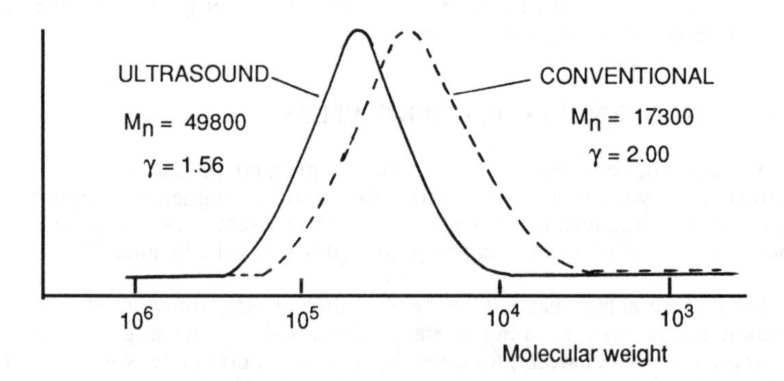

Figure 11. GPC chromatograms of PDMS.

Thus, it seems that ultrasound, under the conditions employed here, cannot be used to "initiate" the ring opening in the absence of added catalyst, there may well be a place for ultrasound in the control of the structure of the polymers arising from the catalysed reaction. Further study of this and other ring opening reactions is underway in our laboratory.

7 POLYMERIZATION EMPLOYING ORGANOMETALLIC REAGENTS

Some of the most useful synthetic applications of 'low molecular weight' sonochemistry have been in the area of heterogeneous reactions involving metal or organometallic surfaces. These reactions are a relatively recent development in polymer chemistry and, with the exception of Ziegler type processes, have yet to reach widespread commercial use. However, a number of reactions have great potential for the preparation of functional polymers and so we have investigated several types.

Ziegler-Natta Polymerizations

The most widespread reaction of this type in polymer synthesis is the Ziegler-Natta process by which materials such as polyethylene and polypropylene are produced industrially on a massive scale[61]. The catalyst is usually a mixed metal species prepared by reacting, for example a trialkyl aluminium with titanium tri- or tetrachloride yielding a complex with a vacant coordination site onto which the alkene is attached. A repeated alkene insertion reaction then results in polymer formation. Although these materials can be produced by other, more straightforward methods, this process produces linear, stereoregular polymers. However, molecular weight control is difficult due to the complexity of the reaction system.

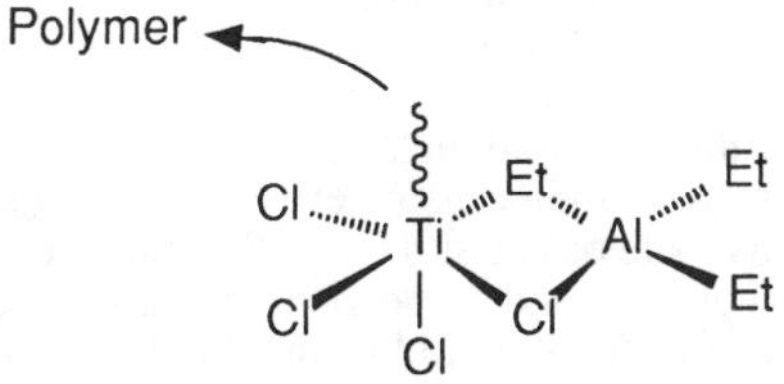

As part of our programme, we have made a preliminary study of the effect of ultrasound on the heterogeneous, Ziegler-Natta polymerization of styrene[62], chosen to give an easily characterizable, model system. A small number of reactions of this type have been carried out with the aim of determining whether ultrasound could influence the rate and yield of the polymerization and also any effect on the molecular weight distribution and microstructure of the polymer.

A $TiCl_4/Et_3Al$ catalyst system was prepared by mixing the reagents in decalin and heating in the usual way[60] under nitrogen gas. Polymerization of styrene was then carried out either in a thermostatted bath or in an ultrasound bath for 18 hours at 30 and 60 °C. The sonicated reactions gave yields of 56% and 24% at 60 and 30 °C respectively while the corresponding values in the absence of ultrasound were 20% and ~5% so that maintaining the sonication throughout the polymerization increased the rate of the reaction. However, it should be stressed that these reactions were not optimized and our yields are rather low (optimization of the stirred reaction should lead to yields in excess of 95%)[60]. Rather, we reproduced reaction conditions (amounts of reactant, temperature, time *etc.*) exactly except for the presence or absence of ultrasound to isolate the effect of the sonication. The NMR spectra of the four polymers showed them to be virtually identical and, as expected, to be almost exclusively isotactic[63,47]. The GPC chromatograms are shown in Figure 12.

The differences caused by sonication here are obvious. The stirred reactions give very broad distributions and it should also be noted that there was a substantial amount of toluene insoluble polymer produced, as is common with this catalyst system[60]. In total contrast, the sonicated reactions gave no toluene insoluble polymer and a narrower, well defined molecular weight distribution. It is possible that, particularly at the higher temperature, some radically initiated, thermal polymerization took place. However, blank experiments showed that any amount would be very small and the ultrasound intensities employed were too low to cause radical formation in any significant amount so that we attribute the results to the effect of the ultrasound.

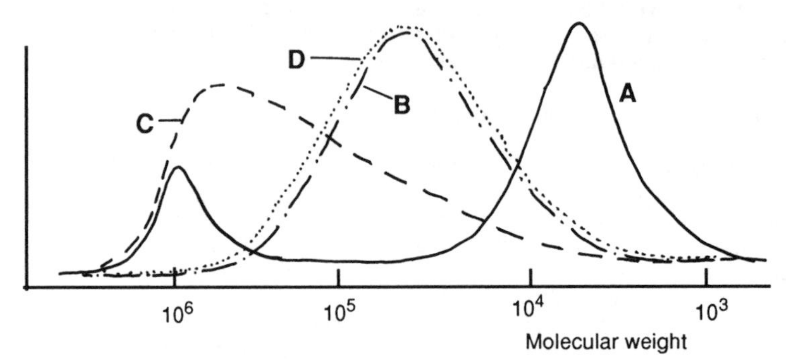

Figure 12. GPC of polystyrenes:
A. Stirred, 30°C; B. Ultrasound, 30°C; C. Stirred, 60°C;
D. Ultrasound, 60°C.

The reasons for the increased yields and rates of reaction are not totally clear but are probably related to sonication causing efficient mixing and faster mass transfer of monomer to the reactive site on the surface of the catalyst. There may possibly also be a reduction in the particle size and consequent increase in active area of the catalyst. One explanation for the usefulness of ultrasound in heterogeneous reactions[42] has been that it has a cleaning action on the surface, resulting in a more homogeneous distribution of active sites. In our case, this would explain, at least in part, the more even distribution of molecular weights found. A second factor leading to a lower polydispersity is the ultrasonic degradation although it would not be expected to be particularly effective at the intensities generated in a cleaning bath.

Polyphenylene materials

These polymers, consisting exclusively of linked aromatic rings, attracted considerable attention some years ago as prototypes for conducting materials. However, their potential has never been fully exploited due to problems in the available synthetic methods[64]. Interestingly, a number of these methods are of the type to which ultrasound has been successfully applied in organic synthesis[65] so that we felt it was worthwhile to explore some of these methods.

The most widely employed method for preparing polyphenylenes has been the oxidative coupling of benzene using a Lewis acid catalyst, the mechanism involving a one electron transfer step[66]. This is precisely the type of reaction that would be expected to be influenced most by ultrasound[65,67] so that we started by carrying out the reaction:

$$C_6H_6 \xrightarrow[AlCl_3\,/\,CuCl_2\;\;H_2O]{40\,^{\circ}C,\;N_2,\;30\;m} \left(C_6H_4\right)_n$$

The conventional method involves reaction in a paste with twice the stoichiometric amount of benzene and sonication of this reaction reduced the yield of polymer, perhaps not surprisingly as a paste is not an ideal medium for ultrasound where a solvent is required to propagate the sound. We therefore repeated the reaction using an ultrasound bath with an excess amount of benzene to give a suspension rather than a paste while keeping the amounts of $AlCl_3$ and $CuCl_2$ the same, the results being shown in Table 3. In each case, elemental analyses were satisfactory for the expected structures and I.R. spectroscopy showed no differences in the ring structure of the materials but ultrasound is clearly accelerating the reaction.

Table 3: Effect of ultrasound on oxidative coupling of benzene.

	VOLUME C_6H_6 (cm^3)	POLYMER YIELD (g)
Stirred	5	0.8
Ultrasound	5	0.4
Stirred	25	0.8
Ultrasound	25	1.9

Other methods for preparing polyphenylenes are based on coupling reactions of dihaloaromatics but they often give low yields and react only as far as the dimer or trimer. We have applied ultrasound to these reactions, again on the basis that they have been some of the most useful sonochemical applications[68] in synthetic organic chemistry. A number of reaction schemes are available but one of the more successful is that due to Rehann *et al.* using a Nickel catalysed Grignard type reaction[69]:

Br–C_6H_4–Br —(Mg, Ni(bipy)Cl_2; N_2, THF)→ –(C_6H_4)$_n$–

Ultrasound was applied to this scheme under various conditions and the results, which show a clear ultrasonic acceleration, are given in Table 4.

In an attempt to simplify the reaction conditions, we then tried to use a Wurtz type reaction using lithium. For example, the coupling of bromobenzene is regarded as a "classic" sonochemical reaction[70,71] giving good yields of biphenyl although the reaction does not work in the absence of ultrasound.

C_6H_5–Br —(Li,)))); THF, 25°C)→ C_6H_5–C_6H_5

Table 4: Effect of ultrasound on Grignard coupling of dibromobenzene.

ULTRASOUND	TEMP. (°C)	TIME (hr)	YIELD (%)
none	Reflux	5	35
bath	60	2	48
none	Reflux	2	20
probe	20	2	33
probe	60	2	45
none	20	14	40

We therefore applied this reaction using 1,4-dibromobenzene in an attempt to produce poly(*p*-phenylene). However, our attempts were not particularly successful. Coupling took place but, based on elemental analyses, only trimers, tetramers and possibly pentamers were produced. Furthermore, I.R. spectroscopy of the solid products suggested that they were not all *para*-substituted rings. These effects are presumably related to the insolubility of the target compounds which, as well as hindering their synthesis, makes them very difficult to characterise since they are also infusible and intractible. One way around this was suggested by Feast *et al.*[72] who prepared polymers substituted with with hexyl- or octyl- groups on the aromatic rings to confer solubility. We have also attempted to use ultrasound in the preparation of these materials but, while it was useful for the preparation of the monomers, also involving Grignard chemistry, we could not achieve polymerization. The starting material here, 2,5-di *n*-hexyl-1,4-dibromobenzene was prepared using a Literature method[72] and we attempted to polymerize this using ultrasound. Use of magnesium with a nickel catalyst as above gave a yield of 48% for a 48 hr reflux in THF (comparable with literature) but a yield of only ~10% after 12 hours sonication on the probe system, showing no significant improvement. We also attempted the reaction by coupling over lithium and several other reactive metals but no significant reaction took place in a Wurtz type reaction.

The lack of reaction in the Wurtz systems prompted further study of the mechanism and we have rationalized our results in terms of a radical type reaction. We studied the Wurtz type coupling of bromobenzene and 2-, 3-, and 4-bromotoluene under varying conditions and have established that it proceeds via a radical mechanism[73,74]. This explains why the coupling is not all *para* and also the reduced reaction in the dialkyl benzenes during polymerization. Thus, although not a success from a synthetic point of view, the work has been useful in extending our knowledge of sonochemical reaction mechanisms.

An alternative approach has been to employ Ullmann reactions utilizing copper in DMF as the coupling agent. Lindley, Lorimer and Mason[75,76] suggested that Ullmann coupling of 2-iodonitrobenzene and 2,5-di bromo nitrobenzene proceeded up to sixty times faster and gave up to 95% yields under the influence of ultrasound. Although we have been unable to duplicate these results, particularly in polymerization systems, we have been able to synthesize new materials. For example, one reaction that we utilized was

$$\text{Br-}C_6(C_6H_{13})_2(NO_2)_2\text{-Br} \xrightarrow{\text{Cu, DMF}} -[C_6(C_6H_{13})_2(NO_2)_2]_n-$$

Under the standard Ullmann conditions of 20 hr reflux at 140 °C in DMF, the reaction gave 95% dehalogenation and negligible polymerization. Conversely, reaction for 4 hr with the ultrasonic probe yielded ~10-15 % of a deep-brown material which we have, as yet, been unable to completely characterize but certainly contains at least four or five rings and we expect to have the structure shown. Similar results were obtained employing nitroiodobiphenyls as "monomers". Although the yields are as yet poor, if it could be increased, this would be a very interesting material since the alkyl groups would confer solubility, the backbone would be potentially conductive while the nitro groups would be very useful to perform further chemical modification to change the properties.

Poly(organosilanes)

These materials, having a backbone exclusively of silicon substituted with a variety of organic groups, are currently attracting considerable interest due to their range of potential applications[77,78]. For example, they are photoactive and photo-conductive and are being investigated for use as photoresists and have also been used as precursors to ceramic materials although they have not fulfilled their early promise in this area.

The usual polymerization method[77-82] uses molten sodium in refluxing toluene in a Wurtz type coupling of dichlorodiorganosilanes. However, the reactions are irreproducible and the yields are rather low, ~55% at best, depending on the nature the substituents. Also, the polymers often have a very wide, usually bi- or tri-modal, molecular weight distribution, usually attributed to two concurrent reaction mechanisms although it has recently also been explained in terms of the solubility characteristics of the components. To achieve commercial use, polymers with a controlled structure and, preferably, monomodal distribution are needed. While these have been produced by carrying out the reaction in the presence of additives such as crown ethers[83,84] a synthetic method to produce them directly would be a significant advance.

The principle of applying ultrasound to this process arose from the discovery in the early 1980's of the facile coupling of chlorosilanes, R_3SiCl, over sonicated lithium metal to give R_3SiSiR_3[85]. The work has been extended to use R_2SiCl_2 to give the polymeric materials.

$$\text{Ph(CH}_3\text{)SiCl}_2 \xrightarrow[\text{Reflux, 110°C / sonicate, 25°C}]{\text{Na, Toluene}} -[\text{Si(Ph)(CH}_3)]_n-$$

We have studied a number of silane "monomers" under a range of conditions[86]. However, from the point of view of commercial exploitation, it seems unlikely that a scheme requiring heating and sonication together would gain widespread acceptance so that we have started our sonochemical reactions at ambient temperature although it should be noted that significant heating can sometimes occur as a result of the sonications.

Dimethyl- and diphenyl-silanes. Reaction of the dimethyl or diphenyl "monomers" produces highly crystalline, insoluble polymers together with oligomeric fractions, soluble in *iso*propanol during work-up. The molecular weight distributions cannot be measured but, as shown in Table 5, both systems show enhanced yields of the high polymer when the sodium is dispersed using ultrasound and the reaction carried out with a cleaning bath.

Table 5. Effect of ultrasound on poly(organosilane) synthesis.

SILANE	CONDITIONS	YIELD (%)	
		Polymer	Oligomer
Diphenyl	Reflux, 110 °C, 5 hr	30	12
Diphenyl	Sonicate, 25-35 °C, 1hr	55	6
Dimethyl	Reflux, 110 °C, 1 hr	12	~70
Dimethyl	Sonicate, 25-35 °C, 1 hr	28	~50

Methyl phenyl silane. We then turned our attention to soluble materials so that we could fully characterize our materials. A number of systems were studied[86] but the poly(methyl phenyl silane) which is soluble in common solvents and so easily characterizable by GPC and NMR, will serve to illustrate our results here.

Ultrasound has been applied to the synthesis of this polymer but with conflicting results. Matyjaszewski and co-workers[87,88] produced materials with monomodal molecular weight distributions, albeit in rather low yield (11-15%) using ultrasound at 60°C in toluene and showed that this only worked for silanes with aryl substituents. Homopolymerization of dialkyl silanes (although they did not use the dimethyl compound) was only possible in more polar solvent systems such as toluene/diglyme mixtures although they could be copolymerized with the aryl compounds in toluene. Conversely, Miller et al.[89] reported that the sonication method did not yield polymers with a monomodal distribution.

We have carried out this polymerization under a wide range of conditions in an attempt to solve this apparent difference. The conventional reflux method in toluene at 110 °C afforded a yield of 15 % after 1 hr. Carrying out the polymerization at 60 °C for the same time in an ultrasound bath and with a "probe" system yielded 35 % and 43 % respectively. The corresponding values at 25 °C were 19 % and 29% respectively. Higher yields could be obtained by allowing the reaction to proceed for longer but these results demonstrate that considerably higher yields can be obtained using ultrasound than by conventional methods.

The effect on the yields is, however, only part of the story. More significant are the changes in the molecular weights and distributions of the polymers. The GPC chromatograms of some of these polymerizations are shown in Figure 13. It should be noted here that the molecular weights given on the chromatograms are relative to polystyrene standards. Clearly, the conventional reflux method gives a polymer with a very wide, bimodal distribution. The polydispersity is vastly reduced under the influence of ultrasound and use of the high intensity probe system gave a monomodal distribution.

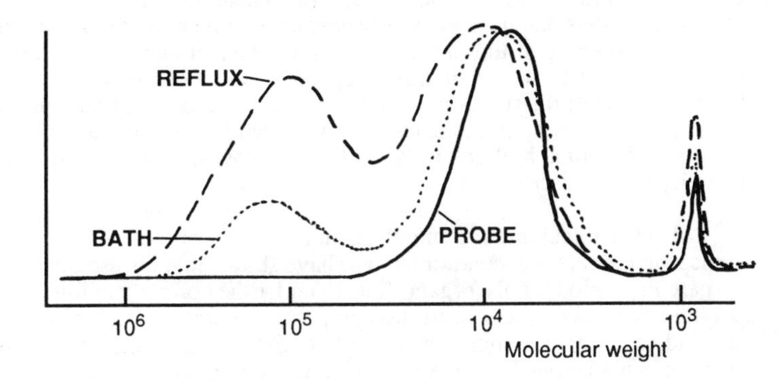

Figure 13. GPC chromatograms of poly(methyl phenyl silane).

It became apparent during our early work that in most cases, the results obtained using the cleaning bath were often very different from those with our ultrasonic probe. This gave a clue as to what may be the explanation for the differences in the published results of Matyjaszewski and Miller. We carried out three polymerizations on the probe system under identical conditions except that the intensity of the ultrasound was varied. The molecular weight distributions of the resulting polymers are shown in Figure 14 and clearly demonstrate that the ultrasound properties play an important part in determining the course of the reaction. The intensities indicated were measured calorimetrically. Similar results have been obtained with several other diorganosilanes. Unfortunately, in common with many, if not most, workers who use ultrasound, neither of the previous authors report the intensity used in their system so that the hypothesis cannot be checked but it seems the most plausible explanation.

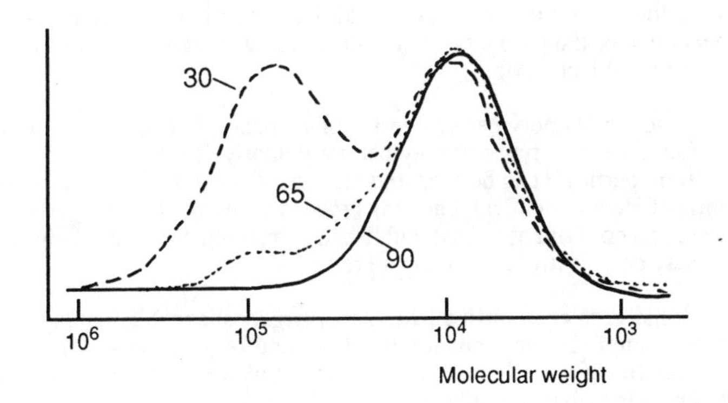

Figure 14. Effect of ultrasound intensity on GPC chromatograms of poly(methyl phenyl silane).
Values indicate ultrasound intensity in W cm^{-2}.

Evidently, the use of higher intensity ultrasound results in much narrower distributions of molecular weight. Many sonochemical reactions have been interpreted in terms of preferential promotion of radical and single electron transfer processes over those involving ionic intermediates. However, there is no firm evidence of radical intermediates in this polymerization and the explanation of the effects probably lies in the physical rather than chemical effects of sonication. The sonochemical acceleration of heterogeneous reactions is usually attributed to increased mass transfer and the continual sweeping of the surface leading to a greater number and faster regeneration of reactive sites. This would, in this case, give a more homogeneous chain growth and hence a narrower distribution of chain lengths. In addition, high molecular weight material formed early in the reaction is degraded by the ultrasound, a process known to be more efficient at high intensity[6,7]. Further work is underway to confirm and quantify these effects.

Thus, we feel that the anomalies in Literature results can be explained by differences in the reaction conditions. We have shown that ultrasound can be used to prepare monomodal poly(organosilanes) under the correct conditions and to prepare polymers with accurately controlled properties. Perhaps more importantly, the results also indicate that care should be taken to optimize and report the precise conditions used for sonochemical reactions if they are to be reproduced by other workers.

8 FUTURE PROSPECTS FOR POLYMER SONOCHEMISTRY

Clearly, the first problem that needs to be addressed before any commercial applications can be considered is the possibility of scaling up the methods to produce significant amounts of material. This is being done by a number of workers in a variety of ways as outlined elsewhere in this Volume so that, should suitable cases arise, the technology either is, or soon will be, available to allow its exploitation. Where then, are these likely to arise?

It seems unlikely that a sonochemical process will replace current methodology unless it allows significant improvement of material properties. Thus, sonochemical manufacture of polystyrene, polyethylene and similar materials on a large industrial scale seems improbable. However, ultrasound has the possibility in a number of areas, for example, radical or emulsion systems, of producing polymers without the need for initiators or emulsifiers etc. There are areas such as biomedical materials and food applications where the presence of additives and impurities is undesirable but where the added value of the products would allow the (at present) extra costs of introducing sonochemical technology.

One area of polymer synthesis not addressed in this review has been electrochemistry. Several types of polymer, particularly those with conducting or non-linear optical properties[90], can be prepared electrochemically. There have been several recent reports of electrochemical reactions proceeding more efficiently under ultrasound[91] and Topare and co-workers[92] have published a preliminary study of electro-polymerization. This may be a fruitful area of future research.

Finally, as exemplified by the poly(organosilanes), a number of organometallic processes are becoming industrially important for polymer synthesis and, with the very clear benefits of sonochemistry in this area, this may well be the major impact of ultrasound in polymer synthesis.

9 CONCLUSIONS

It has been shown that there are a number of areas where ultrasound may be of great benefit in the preparation of polymeric materials. Polymer sonochemistry is a particularly rich field of study as it allows us to utilize all of the various effects of ultrasound. The very high temperatures and pressures generated in cavitation bubbles allow us to produce free radicals to initiate polymerization. The high shear gradients and shock waves around collapsing bubbles cause chain cleavage but may allow us to control the molecular weight of our materials. The very efficient mixing can be employed in emulsion processes and, finally, we can take advantage of the ultrasonic promotion of heterogeneous reactions in a range of organometallic processes.

Several of these synthetic methods are accelerated under sonochemical conditions but perhaps the greatest effects are in the molecular weight distributions of the resulting polymers. Many of the modifications occurring as a result of the shear degradation are understood and can be modelled. It remains to completely characterize the kinetics of polymerization reactions occurring at surfaces before the detailed effects on the polymers of organometallic processes can be understood. However, the outlook is encouraging and it appears that sonochemical polymerization reactions are well placed to become commercially viable in the near future.

10 ACKNOWLEDGEMENTS

It is a great pleasure for me to acknowledge my co-workers who have performed most of the experimental work described in this Chapter: Dr Ali Patel, a postdoctoral research fellow; Paul Smith, Peter West and Andrew Clifton, postgraduate students at Bath; and Melanie Daw, Nick Newcombe and Diane Norris, who did undergraduate research projects in my group. Also, of course, the Science and Engineering Research Council, as well as several industrial collaborators, without whose financial support our work would not be possible.

REFERENCES

1. E.W. Flosdorf and L.A. Chambers *J. Amer. Chem. Soc.* 1933, **55** 3051.
2. A.S. Szalay *Phys. Chem. A.* 1933, **164** 234.
3. A.S. Gyorgi *Nature* 1933, **131** 278.
4. R.A. Pethrick *Advances in Sonochemistry* 1991, **2** 65.
5. B.J. Hunt and S.R. Holding *Size Exclusion Chromatography* Blackie, London, 1989
6. G.J.Price *Advances in Sonochemistry* 1990, **1** 231.
7. A.M.Basedow and K.Ebert *Adv. Polym. Sci.* 1977, **22**, 83.
8. G.J. Price and P.F. Smith *Manuscripts submitted for publication.*
9. B.M.E. Van der Hoff and P.A.R. Glynn *J. Macromol. Sci. Macromol. Chem.* 1974, **A8** 429.
10. W.B. Smith and H.W. Temple *J. Phys. Chem.* 1968, **72** 4613.
11. J.A. Odell and A. Keller *J. Polym. Sci. Polym. Phys.* 1986, **24** 1889.
12. M. Moan and A. Omari *Polymer Degrad. and Stability* 1991, **35** 277.
13. G.J. Price and P.F. Smith *Polym. Int.* 1991, **24**, 159.
14. G.J. Price and P.F. Smith *Eur. Polym. J.* in press.
15. A. Henglein and M. Gutierrez *J. Phys. Chem.* 1988, **92** 3705.
16. G.J. Price et al. *Unpublished results.*
17. A. Noshay and J.E. McGrath *Block copolymers - Overview and critical survey* Academic Press, New York, 1979.
18. H.W. Melville and A. Murray *Trans. Farad. Soc.* 1950, **46** 996.
19. A. Henglein *Makromol. Chem.* 1954, **14** 15.

20. A. Henglein *Makromol. Chem.* 1955, **15** 188.
21. J.R. Ebdon, *New methods of polymer synthesis* Chapman and Hall, London 1991.
22. J.C. Bevington in *Comprehensive Polymer Science* Vol. 3 Ch. 6 J.C. Bevington and G. Allen, (Eds.) Pergamon Press, Oxford, 1989.
23. E. Hart and A. Henglein *J. Phys. Chem.* 1986, **90** 5889, 5992.
24. P. Riesz, Berdahland, D. and Christmoer, C. *Environ. Health. Perspect.* 1985, **64** 233.
25. P. Riesz *Advances in Sonochemistry* 1991, **2** 23.
26. O. Lindstrom and O. Lamm *J. Phys. Colloid Chem.* 1951, **55** 1139.
27. G. Seghal, R.G. Sutherland and R.E. Verall *J. Phys. Chem.* 1982, **86** 2982.
28. *I.E. El'Piner Ultrasound: Physical, chemical and biological effects* Consultants Bureau, New York, 1964.
29. K.F. O'Driscoll and A.U. Shridhan *J. Polym. Sci. Polym. Chem.* 1973, **11**, 1111.
30. H. Fujiwara, H. Kakiuchi, K. Kanmaki and K. Goto *Kobunshi Ronbunshu (Eng. Edn.)* 1976, **5** 256.
31. T. Miyata and F. Nakashio *J. Chem. Eng. Japan* **1975, 8** 463.
32. J.P. Lorimer, T.J. Mason and D. Kershaw, *Ultrasonics International Conference Proceedings* 1989, 1247, 1270.
33. J.P. Lorimer and T.J. Mason, *Ultrasonics International Conference Proceedings* 1987, 762.
34. P. Kruus *Ultrasonics* 1983, **21,** 193.
35. P. Kruus and T.J. Patraboy *J. Phys. Chem.* 1985, **89** 3379.
36. P. Kruus, J. Lawrie and M.L. O'Neill *Ultrasonics* 1988, **26** 352.
37. P. Kruus, M.L. O'Neill, and D. Robertson *Ultrasonics* 1990, **28** 304.
38. P. Kruus *Advances in Sonochemistry* 1991, **2** 1.
39. G.J. Price, M.R. Daw, N.J. Newcombe and P.F. Smith *Br. Polym. J.* 1990, **23** 63.
40. G.J. Price, P.F. Smith and P.J. West *Ultrasonics* 1991 **29** 166.
41. G.J. Price, D.J. Norris and P.J. West *Macromolecules* in press
42. K.S. Suslick *Ultrasound: Its chemical, physical and biological effects*, V.C.H. Publishers, New York, 1990.
43. F.R. Young *Cavitation* McGraw-Hill London 1990.
44. C. Walling and E.R. Briggs *J. Amer. Chem. Soc.* 1946, **68** 1141.
45. J. Brandrup and E.H. Immergut (Eds.) *Polymer Handbook 3rd Ed.*, Wiley, New York, 1990.
46. G.C. Eastmond in *Comprehensive Chemical Kinetics* C. Bamford and C.F. Tipper (Eds) Elsevier, New York, 1976.
47. F.A. Bovey in *Comprehensive Polymer Science* Vol. 1, Ch. 17 Bevington, J.C. and Allen, G. (Eds.) Pergamon Press, Oxford, 1989.
48. F.A. Bovey *Pure. Appl. Chem.* 1966, **12** 525.
49. D.R. Bassett and A.E. Hamielec *Emulsion polymers and emulsion polymerization* A.C.S. Symposium Series **165,** Amer. Chem. Soc., Washington, 1981.
50. Y. Hatate, T. Ikeura, M. Shinonome, K. Kondo and F. Nakashio *J. Chem. Eng. Japan* 1981, **14** 38.
51. Y. Hatate, A. Ikari, K. Kondo and F. Nakashio *Chem. Eng. Commun.* 1985, **34** 325.
52. J.P. Lorimer, T.J. Mason, K. Fiddy, D. Kershaw, R. Groves and D. Dodgson *Ultrasonics International Conference Proceedings* 1989, 1283.
53. K.W. Allen, R.S. Davidson and H.S. Zhang *British Patent Appl. 90177544*, 1990
54.. K.W. Allen, R.S. Davidson and H.S. Zhang *Proceedings of "Radtech Europe" Conference*, Edinburgh 1991.
55. R.S. Davidson *Private communication.*
56. W.J. Bailey in "*Comprehensive Polymer Science*", J.C. Bevington and G.

Allen (Eds.) Pergamon Press 1989. Vol 3, Ch. 22.
57. V. Ragaini *Paper presented to 1st European Sonochemical Society meeting*, Autrans, France 1990.
58. V. Ragaini *Italian Patent Appl.* 20478-A/90.
59. J.J. Leboun and H. Porte in "*Comprehensive Polymer Science*", J.C. Bevington and G. Allen (Eds.) Pergamon Press 1989. Vol 5, Ch. 34.
60. W.R. Sorenson and T.W. Campbell *Preparative methods in polymer chemistry* Wiley Interscience, New York, 1968.
61. P.J.T. Tait in "*Comprehensive Polymer Science*", J.C. Bevington and G. Allen (Eds.) Pergamon Press 1989. Vol 4, Ch. 1.
62. G.J. Price and A.M. Patel *Polymer Commun.* in press
63. N. Ishihara, T. Seimiya and M. Uoi *Macromolecules* 1986, **19** 2464.
64. M.B.Jones and P. Kovacic in "*Comprehensive Polymer Science*", J.C. Bevington and G. Allen (Eds.) Pergamon Press 1989. Vol 5, p. 465.
65. S.V. Ley and C.R. Low "*Ultrasound in Chemistry*" Springer Verlag, London, 1989.
66. P.Kovacic and A.Kyriakis *J. Amer. Chem. Soc.* 1963, **85** 454.
67. J.L. Luche C. Einhorn, J. Einhorn and J. Sinisterrago *Tetrahedron. Lett.* 1990, **31** 4125.
68. T.J. Mason and J.L. Lindley *Chem. Soc. Rev.* 1987. **16**, 275.
69. M. Rehann, A. Schluter, G. Wegner and W.J. Feast *Polymer* 1989, **30**, 1054.
70. T.D. Lash and D. Berry *J. Chem. Educ.* 1985, **62**, 85.
71. B.H. Han and P. Boudjouk *Tetrahedron. Lett.* 1981, **22**, 2757.
72. M. Rehann, A. Schluter and W.J. Feast *Synthesis* 1988, 386.
73. G.J.Price and A.A.Clifton *Tetrahedron. Lett.* 1991, **32** 7133.
74. G.J.Price and A.A.Clifton *Manuscripts in preparation.*
75. J. Lindley, J.P. Lorimer and T.J. Mason *Ultrasonics* 1986, **24** 292.
76. J. Lindley, J.P. Lorimer and T.J. Mason *Ultrasonics* 1987. **25**, 154.
77. R.D. Miller and J. Michl *Chem. Rev.* 1989, **89** 1359.
78. R. West *J. Organometall. Chem.* 1986, **300**, 327.
79. J. Devaux, J. Sledz, F. Schue, L. Giral and H. Naarmann *Eur. Polym. J.* 1989, **25**, 263.
80. R.D. Miller, R. West *et al. J. Polym. Sci. Lett.* 1983, **21**, 819.
81. S. Gauthier and D.J. Worsfold *Macromolecules* 1989, **22** 2213.
82. R.D. Miller *Polym. News.* 1988, **12** 326.
83. M. Fujino and H. Isaka *J. Chem. Soc. Chem. Commun.* 1989, 466.
84. R.H. Cragg, R.G. Jones, A.C. Swain and S.J. Webb *J. Chem. Soc. Chem. Commun.* 1990, 1143.
85. B.H. Han and P. Boudjouk *Tetrahedron. Lett.* 1981, **22**, 3813.
86. G.J. Price and A.M. Patel *Manuscripts in preparation.*
87. H.K. Kim and K. Matyjaszewski *J. Amer Chem. Soc.* 1989, **110**, 3321.
88. K. Matyjaszewski *Polym. Prepr.* 1989, **30(1)** 131; **30(2)** 119
89. R.D. Miller, D. Thompson, R. Sooriyakumaran and G.N. Fickes *J. Polym. Sci. Polym. Chem.* 1991, **29**, 813.
90. H. Sasabe and T. Wada in "*Comprehensive Polymer Science*", J.C. Bevington and G. Allen (Eds.) Pergamon Press 1989. Vol 7, Ch. 6.
91. T.J. Mason, J.P. Lorimer and D.J. Walton *Ultrasonics* 1990, **28** 251.
92. L. Topare, S. Eren and U. Akbulut *Polymer Commun.* 28 36 1987.

Organometallic Processes Promoted by Ultrasound

Philip Boudjouk
DEPARTMENT OF CHEMISTRY, NORTH DAKOTA STATE UNIVERSITY, FARGO, NORTH DAKOTA 58105, USA

1. INTRODUCTION

I want to thank the Royal Society of Chemistry and particularly Gareth Price, the organizer of this symposium, for inviting me to sunny Manchester and giving me this opportunity to present some of our recent results on the effects of ultrasound on heterogeneous reactions.[1] Our speciality is organosilicon chemistry and the work you will see here reflects that focus.

This paper is divided into two sections: I. Stoichiometric Reactions of Group I Metals with Halosilanes, in which our studies of the reactions of dihalosilanes and alkali metals are presented; and II. Transition Metal Catalyzed Reactions of Silanes, which summarizes our results on the reactions of hydrosilanes with platinum on carbon and activated nickel. Both sections illustrate the beneficial effects of ultrasound on reactions involving metals.

2. HISTORICAL BACKGROUND

The first reports of chemical effects of ultrasonic waves by Loomis and coworkers[2] were soon followed by numerous practical applications in which ultrasonic waves were used on heterogeneous systems (Table 1).

Table I Some early practical applications of ultrasonic waves in heterogeneous systems

System	Effects of Ultrasonic Waves	Applications	Ref
Immiscible liquids	More uniform and more stable emulsions; less emulsifying agent required	Processing of foods, drugs, cosmetics, oil recovery	3,4,5
Solid-solid, solid-liquid mixtures	More homogeneous dispersions, higher loading capability	Supported catalysts, alloys, preceramics,	6,7,8
Gas-solid mixtures	Improved mass transport	Drying of coal, fluidized beds, textiles	9,10,11
Electrochemistry	Improved mass transport, reduced overvoltage	Electroplating, fine dispersions	12,13,14

Progress was also made in academic laboratories. Especially noteworthy, in light of the focus of this paper, are Renaud's discovery that the Grignard reaction was accelerated by ultrasound, Pratt's synthesis of phenylsodium and Fry's demonstration that an ultrasonic cleaner could be useful in synthesis. Some important observations involving ultrasound and heterogeneous reactions are given in Table 2.

Table 2 Some heterogeneous reactions accelerated by ultrasonic waves

Reaction			Ref.
$Zn + 2HCl$	))) →	$ZnCl_2 + H_2$	15
R-X + Mg	))) →	R-Mg-X	16
Na + acridine	))) →	$Na_{1.5}^{+}\ (C_{13}H_9N)^{-}$	17
$Na + C_6H_5$-Cl	))) →	$C_6H_5^{-}\ Na^{+}$	18
$CH_3S(O)CH_3$ + NaH	))) →	$CH_3S(O)CH_2^{-}\ Na^{+}$	19
BrR_2C-C(O)-CR_2Br + Hg/HgOAc	))) →	HR_2C-C(O)-CR_2OAc	20

3. STOICHIOMETRIC REACTIONS OF GROUP I METALS WITH HALOSILANES

The recent interest in sonochemistry is due largely to the reports of Luche in late 1980 and to those of Boudjouk and Suslick in early 1981. The motive in our labs was simple: we needed a gentler, more efficient method of preparing metal-metal bonds.

$$R_3M\text{-}X + Li \xrightarrow{)))} R_3M\text{-}MR_3 \quad >85\%$$

M = C, Si, Ge, Sn R = alkyl, aryl, H

Vigorous reaction conditions often lead to oxidation and hydrolysis, not only diminishing yields but also inhibiting purification. For example, disiloxanes, (R_3Si-O-SiR_3), typical byproducts in the reactions of chlorosilanes, are usually difficult to separate from disilanes.

The logical follow-up to these Wurtz type couplings was to examine dihalosilanes.[21] Table 3 shows that cyclization is indeed efficient. An interesting change occurs when the size of the groups on silicon is increased: the ring gets smaller. These are known reactions, and it is also amply documented that, under the right conditions, linear polymeric polysilanes will form in good yield.

The mild conditions of the ultrasound synthesis facilitated a mechanistic study. We wished to trap the intermediate(s) involved in ring formation. Triethylsilane is an excellent trap for carbenes and silylenes. By conducting the lithium reduction of a variety of dichlorosilanes in its presence, we were able to show that, with small groups, silylenes such as Me_2Si:, Et_2Si:, and Ph_2Si: were not produced. In fact, those reactions give essentially the same products in the presence of the triethylsilane as in its absence: cyclic polysilanes. However, when larger groups are placed on silicon, silylene intermediates are indeed generated. And, the larger the groups on silicon, the higher the yield of silylenes (Table 4).

Table 3 Cyclization of dihalosilanes via lithium reduction

Dihalosilane	Major Silacycle Produced
Me_2SiCl_2	$(Me_2Si)_6$
Et_2SiCl_2	$(Et_2Si)_5$
Ph_2SiCl_2	$(Ph_2Si)_4$
$(Mesityl)_2SiCl_2$	$[(Mesityl)_2Si]_3$
$t\text{-}Bu_2SiCl_2$	$(t\text{-}Bu_2Si)_3$

Table 4 Effect of the size of organic substituents on generating silylenes from dihalosilanes

$$RR'SiCl_2 + Li + Et_3Si\text{-}H \xrightarrow{)))} Et_3Si\text{-}SiRR'\text{-}H + RR'HSi\text{-}SiRR'H$$

R	R'	% yields of Et_3Si-SiRR'-H	% yields of RR'HSi-SiRR'H
Me	Me	0	0
Et	Et	0	0
Ph	Ph	0	0
Ph	t-Bu	12	4
Mesityl	t-Bu	18	10
Mesityl	Mesityl	38	0
t-Bu	t-Bu	85	8

The simplest explanation we can offer is that bulky groups inhibit intermolecular coupling providing sufficient time for α-elimination. However, electronic stabilization of the intermediate silylene by aryl groups may also play a role.These trapping reactions provided the first evidence of condensed phase α-elimination reactions of dihalosilanes.

Changing the metal led to a significant change in product distribution (Table 5). An increase in the disilane product at the expense of the silylene insertion product is observed. The source of the hydrogens is not clear, but we can say that most of them do not come from solvent as evidenced by reactions conducted in deuterated solvents. Base abstraction by a silyl anion is more likely, and it is reasonable to suggest that the sodium and potassium ions are more ionic, and consequently more basic, than silyllithium intermediates.

Table 5 Effect of alkali metal on generating di-*tert*-butylsilylene from di-*tert*-butyldichlorosilanes.

$$t\text{-}Bu_2SiCl_2 + M + Et_3Si\text{-}H \xrightarrow{)))} Et_3Si\text{-}Si(t\text{-}Bu)_2\text{-}H + t\text{-}Bu_2HSi\text{-}Si(t\text{-}Bu_2)H$$

M	$Et_3Si\text{-}Si(t\text{-}Bu)_2\text{-}H$	$t\text{-}Bu_2HSi\text{-}Si(t\text{-}Bu_2)H$
Li	85%	8%
Na	51%	21%
K	47%	24%

Varying the halogen produced some surprises (Table 6), the most noteworthy being the apparent inertness of the difluoride to lithium. Certainly this merits more study on our part. We were also surprised that the diiodo compound gave significantly lower yields of insertion product.

Table 6 Effect of halogen substituents on generating di-*tert*-butylsilylene from di-*tert*-butyl-dihalosilanes.

$$t\text{-}Bu_2SiX_2 + Li + Et_3Si\text{-}H \xrightarrow{)))} Et_3Si\text{-}Si(t\text{-}Bu)_2\text{-}H + t\text{-}Bu_2HSi\text{-}Si(t\text{-}Bu_2)H$$

X	$Et_3Si\text{-}Si(t\text{-}Bu)_2\text{-}H$	$t\text{-}Bu_2HSi\text{-}Si(t\text{-}Bu_2)H$
F	0%	0%
Cl	85%	8%
Br	91%	8%
I	66%	11%

The dibromosilane and lithium combination, which gives the highest yield of disilane with triethylsilane, was monitored in a separate study in the absence of trap (Figure 1). We observe that the symmetrical dibromide is formed early followed by the hydrobromosilane and then the dihydro product. Finally, we observe the formation of the novel cyclotetrasilane, 1,3-dihydro-hexa-*tert*-butylcyclotetrasilane. Important to note here is that the major product (80%) of this reaction, hexa-*tert*-butylcyclotrisilane, is not on the plot because, even though it can be isolated by crystallization, it does not survive gas chromatography used for analysis of the reaction products.

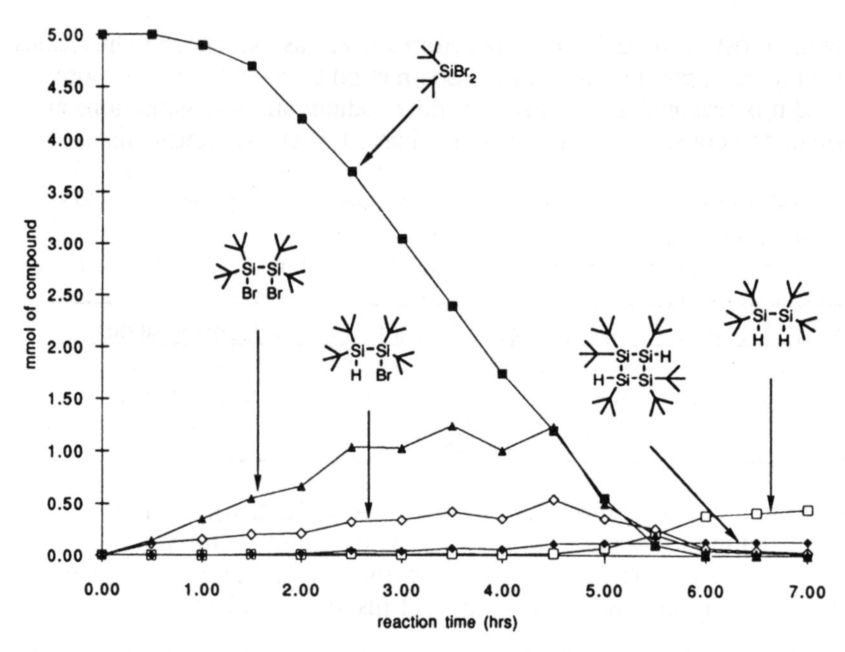

Figure 1. Plot of mmol of reactant and products vs time for the reaction of di-*tert*-butyldibromosilane with lithium in THF in the absence of trap.

Again, taking advantage of the mild conditions, we treated the dichlorosilane with lithium in the presence of olefins (Table 7).[22] For some of these only ultrasound produced the silacyclopropane derivatives. Especially worth noting is that, even though the dibromide had given the highest yields of insertion product with triethylsilane, it is not very efficient in generating silacyclopropanes. Di-*tert*-butyldichlorosilane, on the other hand, is consistently

Table 7 Effects of different leaving groups with various olefin traps

$$t\text{-}Bu_2SiX_2 + Li + \text{Olefin} \xrightarrow{)))} t\text{-}Bu_2Si\text{(silirane)}$$

X	Olefin	Conditions	Yield of Silirane, %
Cl	trans-2-butene	stirring or)))	80
Cl	cyclopentene	)))	55
Cl	cyclohexene	)))	50
Cl	cycloheptene	)))	27
Br	trans-2-butene	stirring or)))	14
Br	cyclohexene	stirring or)))	0

effective with silicon hydride insertion and olefin insertion. Clearly these two halosilanes generate different reactive intermediates.

While only minor differences in product distributions are noted when triethylsilane is used as the trapping agent with the different dihalosilanes, a sharper contrast is noted when olefins are used. Invariably, di-*tert*-butyldichlorosilane gives better yields of silacyclopropanes than either di-*tert*-butyldibromosilane or di-*tert*-butyldiiodosilane. To our surprise, ultrasound was required for these reactions. Simple refluxing gave very low yields.

Some insight into the nature of the silylene generated by di-*tert*-butyldichlorosilane is given in its reaction with *cis* and *trans* butene (Scheme 1). The stereospecific nature of the reaction indicates that the intermediate behaves like a singlet. When treated with bis(trimethylsilyl)acetylene, the intermediate produces high yields of a stable silirene which we have been able to fully characterize.

Scheme 1

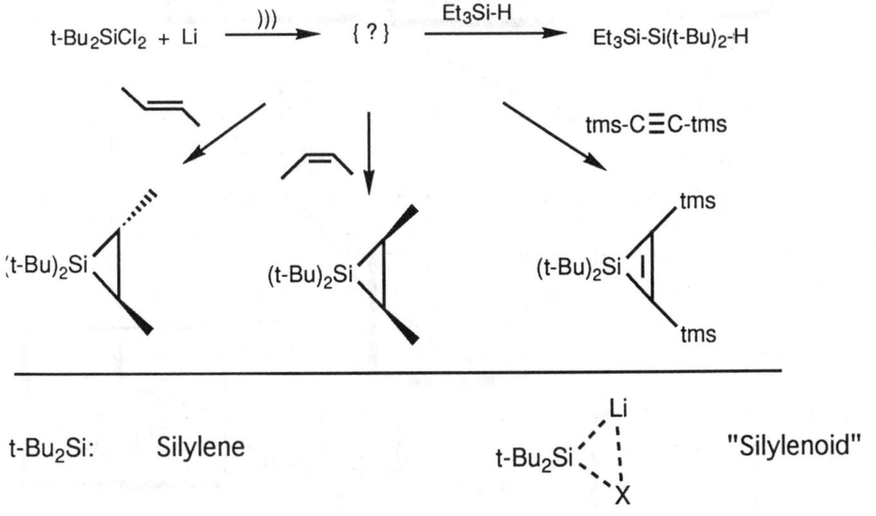

$t\text{-}Bu_2SiCl_2$ + Li))) { ? } $Et_3Si\text{-}H$ $Et_3Si\text{-}Si(t\text{-}Bu)_2\text{-}H$

tms-C≡C-tms

$(t\text{-}Bu)_2Si$ $(t\text{-}Bu)_2Si$ $(t\text{-}Bu)_2Si$ tms tms

$t\text{-}Bu_2Si$: Silylene $t\text{-}Bu_2Si$ Li X "Silylenoid"

These results provide support for the mechanism shown below (Scheme 2) in which the boxed compounds have been unambiguously identified. This is not the final word on this reaction but it is a good basis for discussion.[23]

It has been mentioned that di-*tert*-butyldichlorosilane and di-*tert*-butyldibromosilane differ in reactions with olefins. So, too, when they are treated with lithium in the absence of a trap. These observations lead us to conclude that, at least in some cases, a "silylenoid" is involved. In fact, there may be different kinds of silylenoids generated depending on the leaving groups and the metals employed in the reactions.

Scheme 2

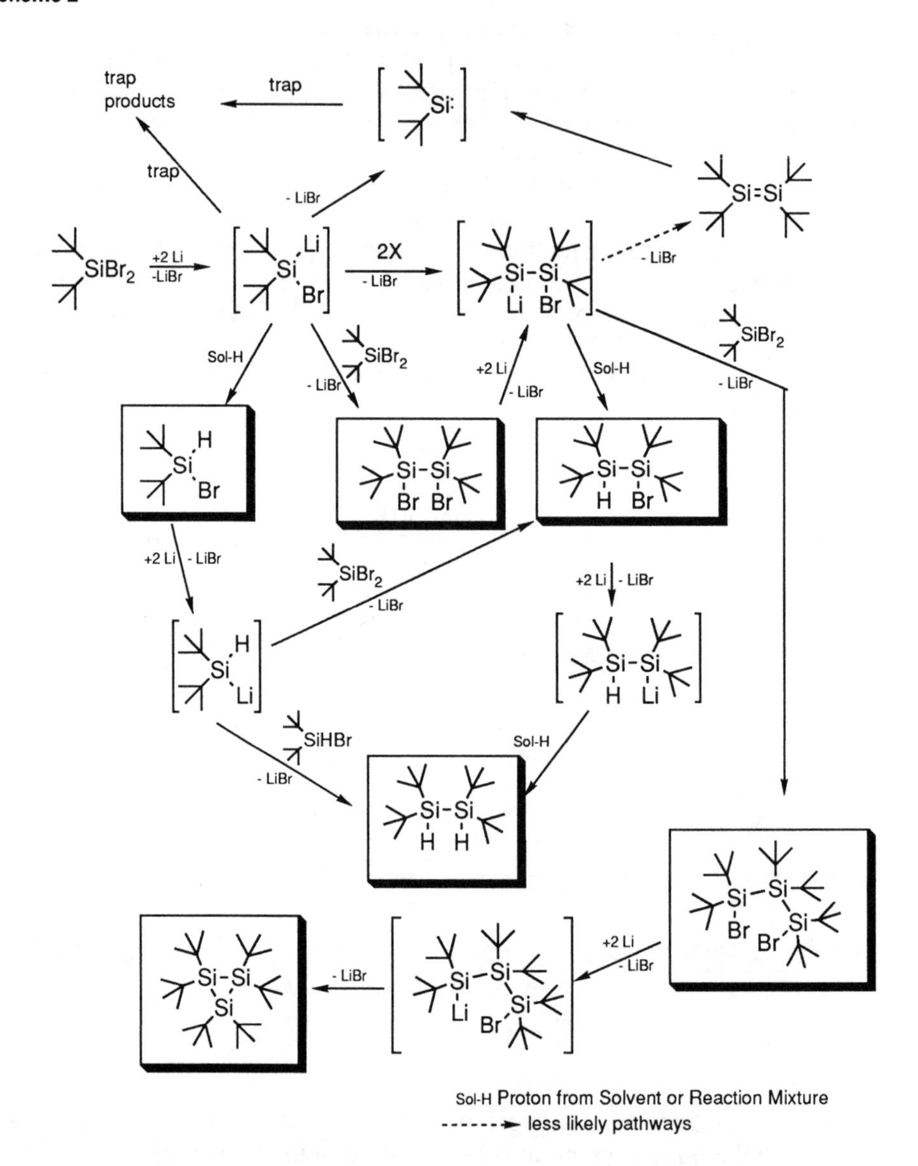

4. TRANSITION METAL CATALYZED REACTIONS OF SILANES

I would like to leave the realm of stoichiometric reactions of the main group metals and move on to those reactions we have studied using ultrasound in which transition metals are used to catalyze the hydrosilation reaction. I should point out that examples of successful applications of ultrasonic waves in catalyzed reactions have been known for several decades. Some examples are listed in Table 8.

Table 8 Early applications of ultrasonic waves in catalysis

System	Effect	Ref.
preparation of platinum from chloroplatinic acid in an ultrasonic field	more active platinum catalyst, with higher surface area and greater magnetic susceptibility	24
preparation of chromium oxide catalyst from chromium nitrate and ammonia	more active catalyst towards H_2O_2 with higher specific area	25
precipitation of calcium and nickel carbonates in an ultrasonic field	increase in activity of 16 -30% towards H_2O_2	26, 27
preparation of iron oxide-chromium oxide catalysit in an ultrasonic field	more active catalyst for oxidation of ammonia	28
preparation of impregnated catalysts in an ultrasonic field (pulsed)	reduced impregnation times, greater activity	29
Haber process in an ultrasonic field	several reports claiming increased rates and yields.	30,31

Our first work in this area involved attempts to improve on the original patent of Wagner and Strother in 1952 which demonstrated that platinum on carbon could catalyze hydrosilation. However, vigorous conditions were employed and the yields were modest.

$$R_3Si\text{-}H + {>}C{=}C{<} + Pt/C \xrightarrow[45\text{ - }115\text{ psi}]{100\text{ - }300^\circ} R_3Si\text{-}C{=}C\text{-}H$$

Our results are shown below indicating a significant improvement over the original reports in terms of yields and conditions. Moreover, we can re-use the catalyst by distilling off the products and recharging the flask.[32]

$$R_3Si\text{-}H + {>}C{=}C{<} + Pt/C \xrightarrow[30^\circ,\ \sim 1h]{)))} R_3Si\text{-}C{=}C\text{-}H$$

R_3 = Cl_3; Cl_2Me; $(EtO)_3$; Et_3

74 - 90%
100% based on conversion

More recently, we examined alkynes.[33] High yields are obtained in all cases. β-Hydrosilylation is the dominant mode of addition for terminal alkynes (Table 9). However, we were surprised to find how efficiently internal alkynes reacted (Table 10). We have determined that only one isomer is produced in each reaction but we do not yet know the stereochemistry.

Because of the expense of platinum, one of our goals is to replace it with a more abundant metal. Nickel seemed a logical choice because of its proximity to platinum, in spite of its poor track record as a hydrosilylation

Table 9 Ultrasonically accelerated hydrosilylation of terminal alkynes

$$RC{\equiv}C\text{-}H + R_3Si\text{-}H + Pt/C \xrightarrow[30^\circ,\ \sim 1h]{)))} \beta\text{-trans} + \beta\text{-cis} + \alpha$$

$RC{\equiv}C$-H	R_3Si-H	Product(s)	Isolated Yields, %
Me_3Si-C≡C-H	Cl_3Si-H	β–trans	92
Me_3Si-C≡C-H	Et_3Si-H	β–trans:β–cis 3:1	88
PhC≡C-H	Cl_3Si-H	β–trans	87
PhC≡C-H	Et_3Si-H	β–trans: α 3:1	91
n-BuC≡C-H	Cl_3Si-H	β–trans: α 11:1	99
n-BuC≡C-H	Et_3Si-H	β–trans: α 19:1	87

Table 10 Hydrosilylation of internal alkynes

$$Et\text{-}C{\equiv}C\text{-}Et + R_3Si\text{-}H + Pt/C \xrightarrow[30^\circ,\ \sim 1h]{)))} (Et)(H)C{=}C(Si)(Et)$$

R_3Si-H	Isolated Yields, %
Cl_3Si-H	95
$MeCl_2Si$-H	82
Ph_2MeSi-H	85
Et_3Si-H	95
$(EtO)_3Si$-H	95

catalyst. We tried nickel in a special form, i.e., nickel powder freshly prepared from nickel halide and lithium in the presence of ultrasonic waves.

$$Li + NiI_2 \xrightarrow{)))} Ni^*$$

Ultrasonic waves accelerate the reduction and produce a more active nickel catalyst than traditional methods of preparation. Remarkably, this nickel catalyzes hydrosilylation very efficiently (Table 11).

Simple and activated olefins react well with trichlorosilane and methyldichlorosilane. The expected terminal product is observed for the former while acrylonitrile gives exclusively the α-adduct. In contrast, vinyl acetate gives only the β adduct in pretty good yield. Methacrylate, on the other hand, gives the O-silated adduct as well as the α-addition product in roughly equal amounts. The β isomer was not detected.

Table 11 Hydrosilylation of olefins catalyzed by activated nickel[34]

Olefin	Silane	Product	Isolated Yield, %
1-hexene	Cl_3SiH	$Cl_3Si(CH_2)_5CH_3$	94
1-hexene	$MeCl_2SiH$	$MeCl_2Si(CH_2)_5CH_3$	92
styrene	Cl_3SiH	$Cl_3SiCH_2CH_2C_6H_5$	88
vinylbutylether	Cl_3SiH	$Cl_3SiCH_2CH_2OC_4H_9$	78
vinylbutylether	$MeCl_2SiH$	$MeCl_2SiCH_2CH_2OC_4H_9$	71
acrylonitrile	Cl_3SiH	$CH_3CH(SiCl_3)CN$	93
acrylonitrile	$MeCl_2SiH$	$CH_3CH(SiCl_2Me)CN$	93
vinylacetate	Cl_3SiH	$Cl_3SiCH_2CH_2O_2CCH_3$	70
methylacrylate	Cl_3SiH	$CH_3CH(CO_2Me)SiCl_3$ + $CH_3CH{=}CO(OMe)SiCl_3$; 1:1	75 - 87

We were surprised to see that, in the absence of an olefin, this activated form of nickel will catalyze the dehydrogenative coupling of silicon hydrides to form Si-Si bonds:[35]

$$2\,Ph_2SiH_2 + Ni^* + PPh_3 \longrightarrow Ph_2HSi\text{-}SiHPh_2 \quad 45\%$$

The above reactions occur only with nickel prepared from the lithium reduction of nickel halides. Ultrasound produces a more active form of this nickel but simple refluxing of the reagents will produce a good catalyst. However, commercial nickel, even after intense sonication was inert toward hydrosilylation and dehydrogenative coupling. Colloidal nickel was ineffective in both reactions as was nickel prepared from the reduction of nickel oxide by hydrogen.

By investigating the surface of freshly prepared nickel powder from lithium and nickel iodide using Raman spectroscopy, we observed peaks at 1956 and 2051 cm^{-1}, diagnostic for bridging and terminal metal carbonyls, respectively[36] (Figure 2). Remarkably, only nickel powder with these bands were active as hydrosilylation catalysts. Removal of the carbonyls by hydrogen at 200 °C completely deactivated the nickel.

Subsequent studies in our laboratories reveal the probable source of the carbonyl groups is the solvent. Apparently the nickel is so active that it can breakdown tetrahydrofuran, diethyl ether and dimethoxyethane and abstract a carbon and an oxygen and form nickel carbonyl linkages.

Addition of hydrosilanes to the activated nickel powders generated a new band in the Raman spectrum at 1947 cm^{-1} (Figure 3). On the basis of studies using Si-D analogs and observing a Ni-D absorption, the 1947 cm^{-1} band was assigned to the Ni-H stretch. This results, presumably, from the oxidative addition of the Si-H bond to nickel. A simple mechanism for the reaction is given below (Scheme 3). However, the role of the carbonyls in promoting these new catalytic reactions of nickel, although clearly essential, is still undetermined.

Studies of the surface of the nickel powder, prepared by lithium reduction of nickel iodide in an ultrasonic bath, show morphology not observed on nickel powder prepared in the absence of ultrasonic waves. As Figure 4 illustrates, there are craters on the surface of the nickel. It seems particularly noteworthy that the craters are smooth, even under high magnification, obviously not the result of simple pitting. We can speculate, based on the known ability of cavitating ultrasonic waves to generate temperatures in excess of 2000 °C, that portions of the nickel surface may become hot enough either to melt, or perhaps even to boil, and that the smooth craters may be the result of plastic deformations from the shock waves or, the bursting of bubbles of molten nickel.

Scheme 3

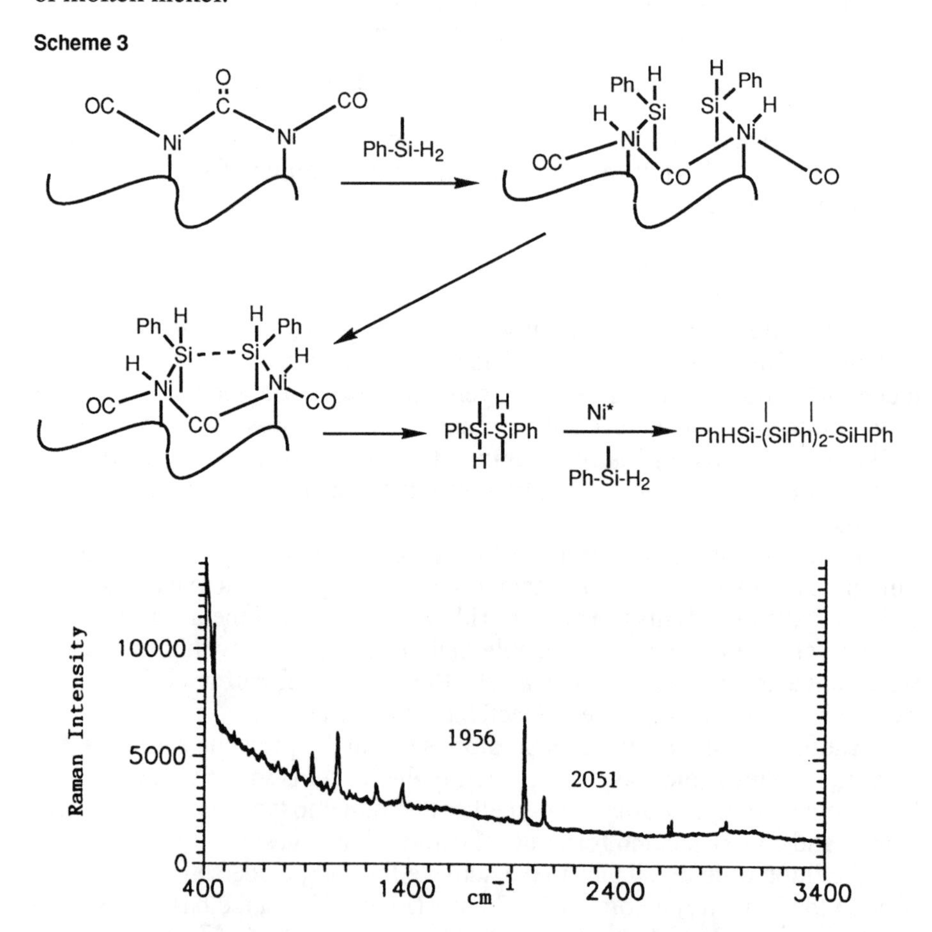

Figure 2. Raman Spectrum of activated nickel powder formed by ultrasound promoted reduction of NiI_2 with lithium metal. Tetrahydrofuran solvent was removed under vacuum.

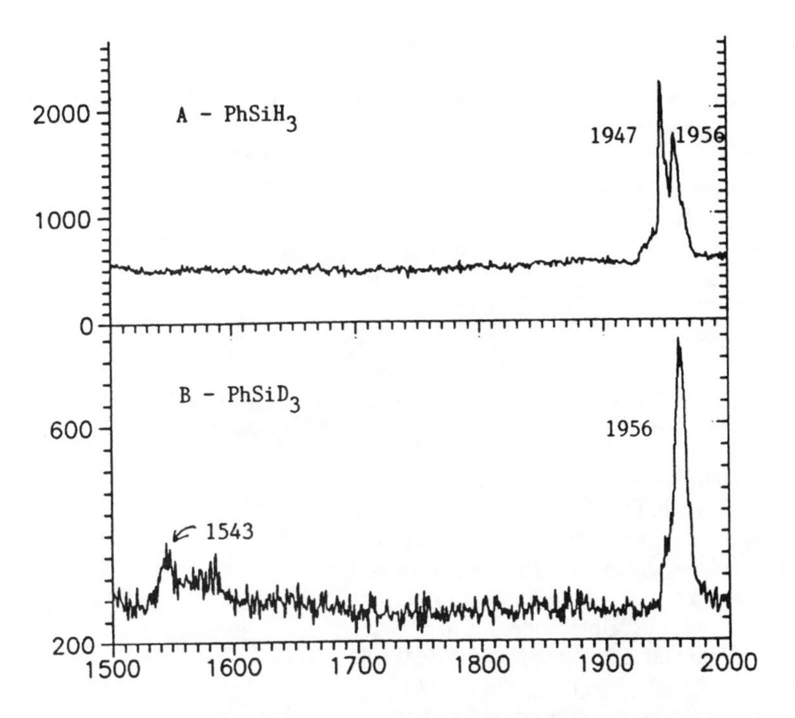

Figure 3. Raman spectra of phenylsilane added to a suspension of activated nickel in THF.

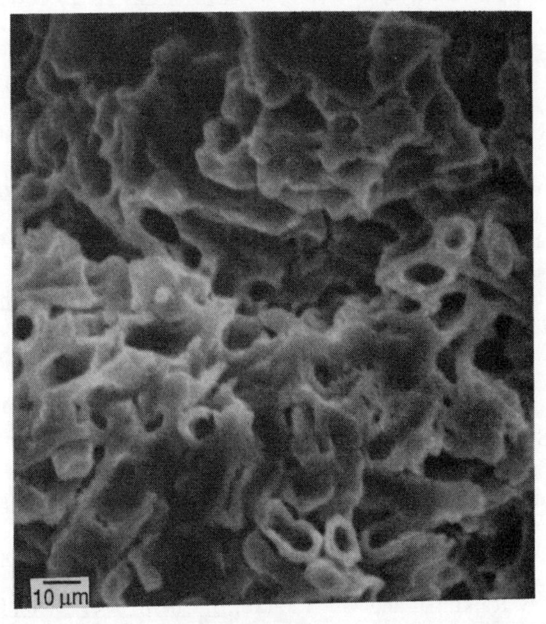

Figure 4. Scanning electrom micrograph of activated nickel powder formed by ultrasound promoted reduction of NiI_2 with lithium metal.

REFERENCES

[1] P. Boudjouk, in 'Ultrasound: Chemical, Physical and Biological Effects', K. S. Suslick Ed., Verlag Chemie International, 1988, Chapter 5, p. 165.

[2] W. T. Richards and A. L. Loomis, J. Amer. Chem. Soc., 1927, 49, 3086.

[3] Siemens, Schuckertwerke Akt.-Ges., Swiss Pat. No 294,390; 1944.

[4] R. Decker and H. Holz, German Pat. No. 803,538; 1951

[5] C. A. Komar and H. A. W. Moore, Chem. Eng. Sym. Ser., 1971, 67 117.

[6] K. Söllner, Trans. Faraday Soc., 1938, 34, 1170.

[7] M. A. Margulis, Russian Pat. 631,372; 1959.

[8] E. A. Heidemann, J. Acoust. Soc. Am. 1954, 26, 831.

[9] R. S. Soloff, J. Acoust. Soc. Am. 1964, 35, 961.

[10] J. S. Wilson, A. S. Moore and W. S. Bowie, Chem. Eng. Sym. Ser., 1971, 67, 68.

[11] K. R. Purdy, G. W. Simmons, A. E. Hribar and E. I. Griggs, Chem. Eng. Sym. Ser., 1971, 67, 55.

[12] N. Moriguchi, J. Chem. Soc. Jpn., 1934, 55, 749.

[13] B. Claus, Z. Tech. Physik, 1935, 16, 80.

[14] F. Mueller and H. Kuss, Helv. Chem. Acta 1950, 33, 217.

[15] N. Moriguchi, J. Chem. Soc. Jpn., 1933, 54, 949.

[16] P. Renaud, Bull. soc. chim. (France), 1950, 1044.

[17] W. Slough and A. R. Ubbelohde, J. Chem. Soc., Chem. Commun. 1951, 918.

[18] M. W. T. Pratt and R. Helsby, Nature, 1959, 184, 1694.

[19] K. Sjöberg, Tetrahedron Lett., 1966, 6383.

[20] A. J. Fry and D. Herr, Tetrahedron Lett., 1978, 1721.

[21] P. Boudjouk, U. Samaraweera, R. Sooriyakumaran, J . Chrusciel and K.R. Anderson, Angew. Chem., Intl. Ed., 1988, 27, 1355.

[22] P. Boudjouk, E. Black and R. Kumarathasan, Organometallics, 1991, 10, 2095.

[23] P. Boudjouk, R. Kumarathasan, U. Samaraweera, E. Black, S. Castellino, J. P. Oliver and J. W. Kampf, submitted for publication.

[24] A. N. Mal'tsev, I. V. Solov'eva, Zh. Fiz. Khim.,1970, 44, 1092. Chem Abstr. 1970, 73, 39,022r.

[25] E. Kowalska, M. Misczyszn, Rocz. Chem., 1972, 46, 233. Chem. Abstr. 1972, 76, 145,322k.

[26] Z. Bronislaw and A. Lomnicka, Zesz. Nauk. Univ. Jagiellon; Pr. Chem., 1975, 20, 109. Chem. Abstr., 1976, 85:25765w.

[27] S. Witekowa and W. Forbotko, Soc. Sci. Lodz Acta Chim., 1974, 18, 99403. Chem. Abstr. 1975, 82:7935k.

[28] A. V. Romenskii, I. V. Popok, and A. Ya. Loboiko, Atroshchenko, Khim. Technol., 1985, 1, 21. Chem. Abstr. 1985, 102: 120614z.

[29] W. Marti and K. Waldemeir, Swiss Pat. No. 376,886, 1964. Chem. Abstr. 1964, 61:P10,088.

[30] C. N.Richardson, U. S. Pat. 2,500,008, 1950. Chem. Abstr. 1950, 44, P4643.

[31] J. B. Jones, U. S. Pat. 3,245,892, 1966. Chem. Abstr. 1966, 65, P82f.

[32] B-H. Han and P. Boudjouk, Organometallics, 1983, 2, 769.

[33] P. Boudjouk and B. Hauck, XXV Organosilicon Symposium, Los Angeles, April 1992, Abstract No. 82.

[34] P. Boudjouk, B.-H. Han, J. R. Jacobsen, and B. J. Hauck, J. Chem. Soc., Chem. Commun., 1991, 1424.

[35] P. Boudjouk, A. B. Rajkumar and W. L. Parker, J. Chem. Soc. Chem. Commun. 1991, 245.

[36] W. L. Parker, P. Boudjouk and A. B. Rajkumar, J. Am. Chem. Soc., 1991, 113, 2785.

Ultrasound in Catalytic and Solid Supported Reagent Reactions

James Lindley
SCHOOL OF APPLIED CHEMISTRY, COVENTRY UNIVERSITY, PRIORY STREET, COVENTRY CV1 5FB, UK

INTRODUCTION

There are two important effects when ultrasonic waves pass through a fluid medium. Firstly, there is enhanced molecular motion caused by acoustic streaming, which arises from non-linear coupling of the first order acoustic waves and gives rise to the well known stirring effect of ultrasound. In the vicinity of surfaces ultrasonic streaming gives rise to strong convective currents which cause a reduction in the thickness of diffusion layers thereby enhancing processes which are mass transport controlled. Secondly, at sufficiently high acoustic intensities ultrasound produces cavitation in liquids. Cavitation is the process in which microbubbles, which are formed within a liquid during the rarefaction phase of the acoustic wave, undergo violent collapse during the compressive phase of the wave. During the compressive phase the bubble contents are adiabatically heated to temperatures estimated to be in the region of 5000 K, and the surrounding thin liquid shell is heated to around 1900 K. These liquid phase 'hot spots' are estimated to have radii around 200 nm and lifetimes of less than 2 μs,[1] although recent sonoluminescence experiments suggest possible lifetimes of <1 ns.[2] Additionally, the implosion of the cavitation bubbles generates high energy shock waves with pressures of several thousand atmospheres. In the vicinity of surfaces bubble collapse is non-spherical and high velocity microjets of solvent are produced which impinge on the surface with velocities estimated to be as high as 100 ms^{-1}.[3]

These effects may lead to improved reactivity of solid reagents by:

(i) removal of passivating surface coatings,
(ii) creation of surface defects, which because of reduced coordination may lead to reactive centres,
(iii) reduction of particle size thereby increasing the reactive surface area,
(iv) improvements in mass transport.

For heterogeneous reactions involving solid catalysts/ reagents ultrasound may be beneficial in three distinct phases:

(i) in catalyst/reagent preparation,
(ii) in catalyst/reagent activation,
(iii) during reaction with substrate.

ULTRASOUND IN CATALYST/REAGENT PREPARATION

Crystallization

The use of ultrasound to enhance rates of crystallization was reported as long ago as 1927.[4] Since then there has been considerable increase in activity in this field particularly in the area of crystallization of metals and alloys.[5] With acoustic fields below the cavitation level acoustic streaming is effective in increasing mass transport of crystal building material to growing crystals which can lead to an increased rate of crystal growth provided that the degree of supersaturation is low. At high degrees of supersaturation, the concentration gradient near to the growing crystal is also high so that an intensive flow of construction material to the crystal is available and acoustic streaming has little effect on crystal growth rate. At acoustic intensities above the cavitation level increases in both the rate of nucleation and crystallization may be observed; this often results in crystals with a smaller particle size and narrower particle size distribution. The crystal habit of ultrasonically produced crystals is often different to conventionally produced crystals as they tend to show less dendritic growth, which gives rise to crystals with a more equiaxed form. Although the mechanism of these effects is not fully understood one proposal is that during the expansion cycle of the acoustic wave the bubble contents are adiabatically expanded which gives rise to a localised cooling in the vicinity of the bubble and an increase in the degree of supersaturation, this leads to the formation of germ nuclei which are dispersed throughout the medium following the ensuing bubble collapse.[6]

We have recently been investigating the effects of ultrasound in zeolite synthesis.[7] Zeolites are microporous tectosilicates with giant 3-dimensional lattices built up from AlO_4 and SiO_4 tetrahedra. Zeolites are usually prepared hydrothermally by heating aqueous solutions of sodium silicate and sodium aluminate at temperatures from 25 to 300°C for several days to a few hours. We found that in the synthesis of zeolite NaA ultrasound led to substantial reductions in nucleation time and overall completion times compared with control reactions. *Table 1.*

Table 1. Synthesis of Zeolite NaA at 85°C

Ultrasonic system	Seeded S Unseeded U	Nucleation time(h)	Completion time(h)
Cleaning bath (50kHz,150W)	U	3	7
Control	U	4	9
Cleaning bath	S	0.75	5
Control	S	2.5	7
Probe[a]	U	1	3.5
Control	U	5	10

a. Sonics and Materials, Vibracell System, 500W.

The progress of the reactions was followed by X-ray diffraction (*Figure 1)* and water absorption (*Figure 2)* at constant relative humidity. The aluminate and silicate solutions produce an amorphous gel on mixing which after a nucleation period rapidly crystallizes to give the cubic NaA crystals.(*Figure* 3).The ultrasonically produced crystals showed reduced particle size over the control. (*Figure 4*). In separate experiments it was demonstrated that the particle size reduction was not a result of interparticle collisions, since a reaction which was carried out to completion in the conventional way and then subjected to ultrasound of the same power used in the ultrasonic syntheses showed no further reduction in particle size. Additionally, crystallization in the presence of seeds also led to particle size reductions in the presence of ultrasound, which seems to support a liquid phase nucleation process in which ultrasound aids the dissolution of the seeds to yield germ nuclei which are dispersed throughout the medium as a result of cavitation.

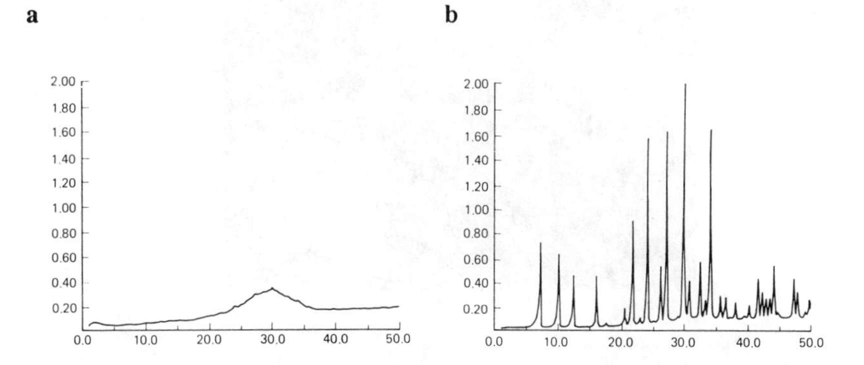

Figure 1 X-ray diffraction patterns during zeolite NaA synthesis (a) 4 h and (b) 12 h.[7]

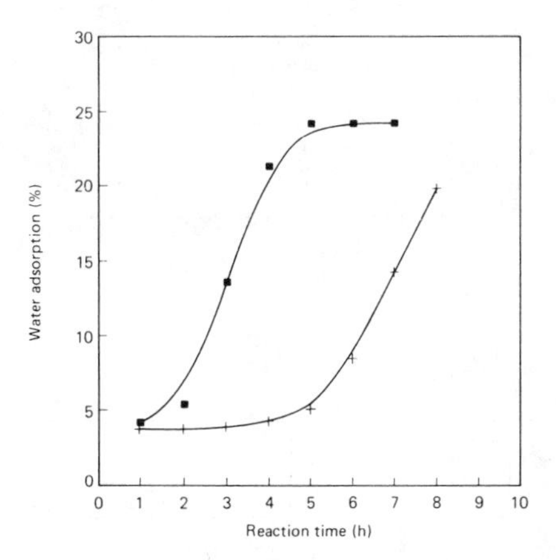

Figure 2 Water adsorption during zeolite NaA synthesis at 75% relative humidity and 85°C. ■ continuous ultrasound (Sonics and Materials, Vibracell System, 500 W)[7]

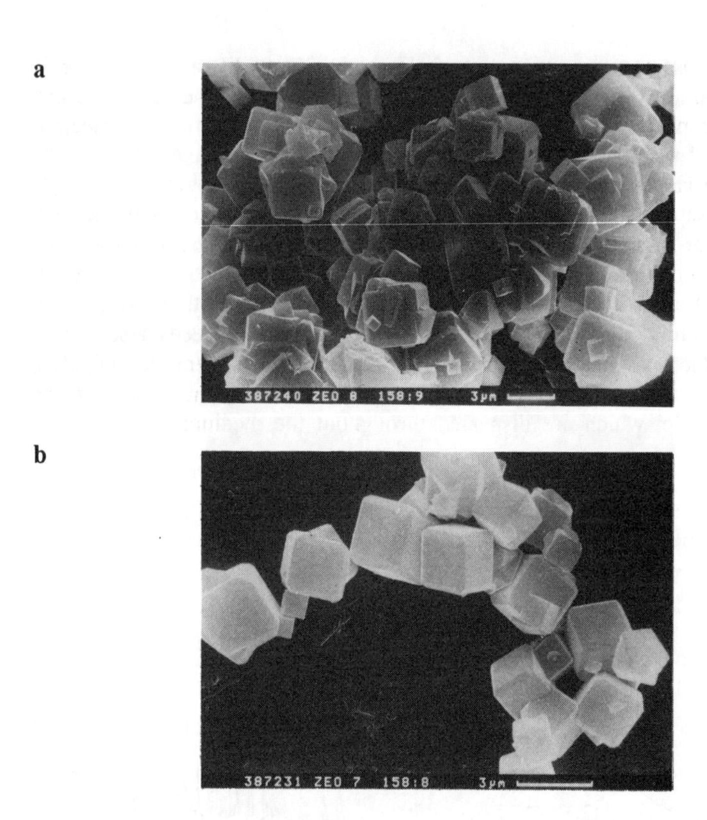

Figure 3 Scanning electron micrographs of zeolite NaA. (a) Ultrasound and (b) mechanical stirring.[7]

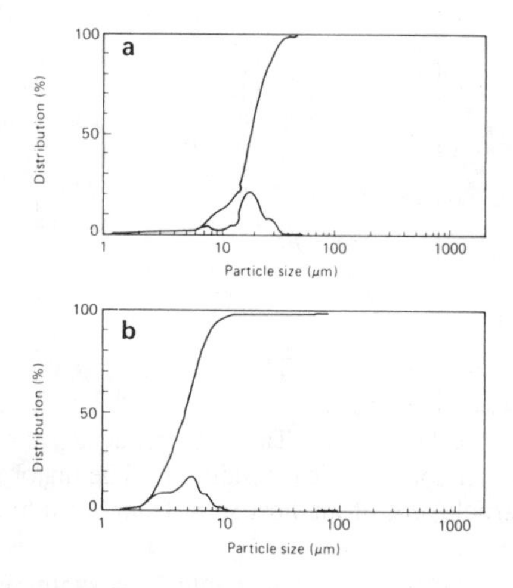

Figure 4 Particle size distribution of zeolite NaA. (a) Mechanical stirring and (b) ultrasound.

Ultrasound has also been reported to be of value in the hydrothermal synthesis of A-type zeolites by enabling cheap natural minerals such as kaolin to be used as the aluminosilicate source.[8]

The use of ultrasound during the precipitation of a $Cs_2Pb_{0.2}PMo_{12}O_x$ catalyst for the gas phase oxidation of methacrolein and a $FeTe_{0.85}MoO_x$ catalyst for the oxidation of alkenes from aqueous solution gave catalysts with increased specific surface area which led to increased catalytic activity.[9] Similar results have been reported for the precipitation of a mixed chromium-molybdenum catalyst for the oxidation of methanol to formaldehyde.[10] However, application of the ultrasound after the precipitation was complete led to agglomeration, surface area reduction and loss of activity.[11]

Amorphous Metals

Platinum metal (Pt, Pd, Rh) blacks prepared by reduction of aqueous solutions of soluble salts in the presence of low intensity ultrasound show up to 62% increase in surface area and increased activity in the decomposition of hydrogen peroxide compared with non-ultrasonic controls.[12]

Recently, a sonochemical method for the synthesis of amorphous iron has been described.[13] In this method $Fe(CO)_5$ in decane is irradiated at 0°C with high intensity ultrasound(20 kHz, 100 Wcm^{-2}) for 3h under argon. The amorphous nature of the iron was confirmed by X-ray powder diffraction, differential scanning calorimetry and electron microscopy. The amorphous iron produced by this method showed ten times greater activity in the catalysis of the Fischer-Tropsch conversion of carbon monoxide and hydrogen to low molecular weight hydrocarbons at the low reaction temperature of 200°C and >30 times activity in the dehydrogenation of cyclohexane to benzene compared with 5 μm crystalline iron powder. The production of amorphous iron is critically dependent on the energetics of cavitation which is strongly influenced by solvent vapour pressure, thermal conductivity of dissolved gas and the ratio of specific heats of the dissolved gas. Thus sonolysis of $Fe(CO)_5$ in a relatively high vapour pressure solvent such as pentane instead of decane leads mainly to cyclotrimerisation to $Fe_3(CO)_{12}$. The formation of amorphous iron involves multiple dissociation of CO ligands within the cavitation bubble to give iron atoms, which have been detected by the presence of iron atomic emission lines in the sonoluminescence spectra.

Intercalation

Spectacular improvements in the rates of intercalation of guest molecules into layered inorganic solids have been reported by Green and Suslick:[14] thus the time for intercalation of n-hexylamine into TaS_2 was reduced from 50h under thermal conditions to 0.25h under sonochemical conditions. The major sonochemical effect was a reduction in particle size of the host solid from 75 to 5μm. In our group we have observed similar effects in the rates of intercalation of metal ions into clays; for example, the time for equilibration during the exchange of copper nitrate into montmorillonite and hectorite clays was reduced from 48h in the conventional shaking method to only 1h with an ultrasonic cleaning bath.[7](*Table 2*)

Table 2. Intercalation of Copper(II) Nitrate into Hectorite

Concentration	0.1M $Cu(NO_3)_2$		1.0M $Cu(NO_3)_2$	
Method	US[a]	Mechanical shaking	US	Mechanical shaking
ESR[b]				
$g_{parallel}$	2.243	2.275	2.264	2.268
$g_{perpend.}$	2.010	2.015	2.015	2.016
Lattice d spacing/nm	125	128	127	131
Copper exchanged (mg/g)	3.49	3.36	5.98	5.11

a. US cleaning bath, 150W, 50kHz
b. Anisotropic ESR signals are characteristic of intercalated Cu^{2+}

The copper exchanged montmorillonite produced by the ultrasonic method showed reduced particle size and increased activity in the catalysis of the decomposition of hydrogen peroxide compared with a control.

Impregnation of Catalysts and Reagents on Supports

The last decade has seen spectacular growth of interest in the use of catalysts and reagents impregnated on high surface area supports such as silica gel, kieselguhr, molecular sieves, alumina, celite.[15] The use of such systems has not only led to substantial rate and yield enhancements but also has the advantage in many cases of simple work-up procedures. In several cases the combined use of ultrasound and support has led to improvements in reactivity of a number of inorganic reagents/catalysts for organic reactions.[16]

The use of ultrasound to facilitate the impregnation of reagents on supports is not new: the first report appeared in the patent literature in 1964.[17] Subsequently Ranganathan *et al*,[18] reported that sonication for 1h of aqueous suspensions of various metal oxides (Cr_2O_3, MnO_2, Co_2O_3) and alumina prior to drying and calcination yielded catalysts with higher dispersity of metal oxide and higher activity in the decomposition of hydrogen peroxide. Similarly, reduction of aqueous ammonium hexachloroplatinate solution containing a suspension of silica gel in an ultrasonic field (440 kHz, 5 Wcm^{-2}) gave a 80% increase in surface area of the metal compared with a mechanically stirred control.[19]

In our group we have recently been investigating the effect of ultrasound on the bromination of aromatic hydrocarbons using copper(II) bromide on neutral alumina.[7] The use of copper(II) chlorides and bromides as halogenating agents for aromatic hydrocarbons was first proposed by Kochi;[20] later studies by Nonhebel,[21] and independently by Tanimoto,[22] found that only aromatic hydrocarbons with ionisation potentials <7.55 eV could be halogenated even under quite forcing conditions. Recently, Kodomari reported that copper(II) chloride or bromide supported on alumina allowed the halogenation of aromatic hydrocarbons with ionisation potentials above 7.55 eV under relatively mild conditions.[23]

Our results for the bromination of naphthalene are shown in *Table 3.* For consistent results and optimum yields a mole ratio of $CuBr_2$ to naphthalene of 5 to 1 was found to be necessary. The reaction is also sensitive to the presence of water; the highest rates and yields were obtained when 0.04 mole of water per mole of copper was added to the dried reagent. Reaction without preimpregnation of the reagent on the support (run 1) was subject to a 2 h induction period which demonstrates the synergy which exists between the reagent and support. Entries 3 and 4 show that ultrasound leads to substantial rate improvements over the Kodomari method. However, the use of ultrasound during the impregnation of the $CuBr_2$ on the support, entry 5, followed by reaction under nonsonochemical conditions, gave the best results. The reagent prepared in this way was shown to have reduced particle size and that the ultrasound had led to a considerable change in the surface morphology.(*Figure 5*) Studies of the effects of the variation of the ultrasonic power used in the preparation of the $CuBr_2$/alumina reagent on reactivity in the bromination of naphthalene revealed that increasing the ultrasonic power up to an optimum level led to increasing surface disruption which was accompanied by increased reactivity. Further increases in power led to agglomeration and decreased reactivity.

Table 3. Bromination of naphthalene using $CuBr_2/Al_2O_3$ at 76°C in CCl_4

Run	Reagent Preparation	US irrad	Rel. init. rate	100% conv. time[a]	% yield[b]	
					1-Br	2-Br
1	$CuBr_2$ and Al_2O_3 without prior mixing	No	0.33[c]	--	55	--
2	$CuBr_2/Al_2O_3$ after Kodomari[23]	No	1.00	180	91	9
3	as in run 2	Yes	1.40	100	89	11
4	as in run 2	Yes[d]	1.80	100	86	14
5	Aqueous suspension of $CuBr_2$ and Al_2O_3 insonated for 1h then as in run 2	No	1.80	60	87	13

a. Minutes
b. Yields calculated from glc data
c. Calculated after a 2h induction period
d. In chlorobenzene

The brominations are inhibited in the presence of radical scavengers such as DPPH and dinitrobenzene. The mechanism of the bromination is thought to involve the formation of aromatic radical cations by single electron transfer from the hydrocarbon to the $CuBr_2$, this process is known to be facilitated by neutral alumina.[24] The use of supports such as montmorillonite, silica gel and celite gave inferior yields.

a

b

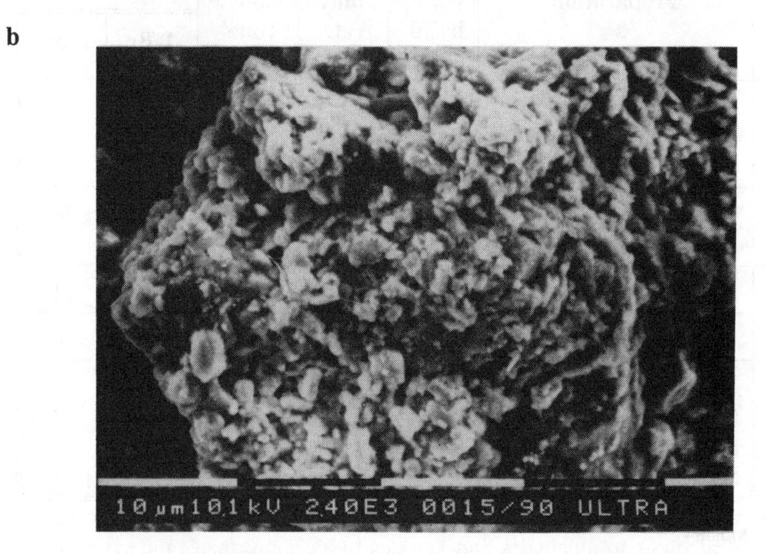

Figure 5 Scanning electron micrograph of $CuBr_2/Al_2O_3$ (a) Mechanical stirring and (b) ultrasound.[7]

ULTRASOUND IN CATALYST/REAGENT ACTIVATION

The extreme turbulence within liquids caused by cavitation has considerable effects on the surfaces of solid particles suspended within the liquid. These changes may arise as a result of bombardment of the surface by solvent microjets and shockwaves or as a result of high energy interparticle collisions driven by the cavitational forces. In some cases the interparticle collisions between metal particles are of sufficient energy to cause local melting and fusion of the colliding particles.[25] In the case of hard metallic reagents /catalysts such as copper and nickel powders ultrasonic pretreatment causes the removal of passivating oxide coatings on the surface. *Figure 6* shows scanning electron micrographs of copper powder which after sonication for 1 h undergoes a particle size reduction from 86 μm (6a) to 23 μm (6b) and the original rough surface (6c) has become smooth (6d). This sonicated copper shows enhanced activity in the Ullmann coupling of activated aryl halides.[26] Similarly, ordinary 5μm nickel powder, which has very low activity as a catalyst for alkene hydrogenation, becomes activated after sonication for 1 h in pentane. Thus the hydrogenation of 1-nonene to nonane is catalysed at a rate of 1.5 mM/min using presonicated nickel, compared with only <10 nM/min with the unsonicated nickel. The rate of hydrogenation with the sonicated nickel is comparable to that of active forms of nickel such as Raney nickel.[27] Raney nickel itself is reported to have enhanced activity in the stereospecific deuteration of carbohydrates after sonication. Extensive surface studies by XPS, Auger electron spectroscopy and SIMS shows that the acoustic field increases and develops the catalytic sites, removes passivating impurities and causes an elemental redistribution within the bulk catalyst.[28, 29] There are many reports in the patent literature of the use of ultrasound in the regeneration of metallic catalysts.[30]

The activation of non-metallic inorganic catalysts/reagents by ultrasound is well documented.[16] The technique ranks alongside other methods of activation such as phase transfer catalysis and the use of high surface area supports. In many cases the combination of ultrasound and the other techniques gives enhanced results. For example, the combination of phase transfer catalysis and ultrasound has been successfully exploited in both liquid-liquid and liquid-solid phase systems.[31] The results illustrated in *Table 3* for the bromination of naphthalene using $CuBr_2$ on alumina illustrate the combined effects of ultrasound and high surface area supports. Ando[16] has shown that for many reactions promoted by non-metallic inorganic solids it is necessary to have a small quantity of water present for optimum reactivity and in many cases this water is provided by the 'inert' support. Furthermore, the optimum number of moles of water is related to the lattice energy of the inorganic reagent and that the function of the water is to activate the surface of the solid reagent by removal of some of the surface ions by solvation. Using the hypothesis that water was necessary for surface disruption, Ando argued that ultrasound may fulfil the same role. This argument led to the identification of several successful sonochemical syntheses such as the oxidation of alcohols to aldehydes and ketones by permanganate, the synthesis of acyl cyanides, and the synthesis of aminonitriles as precursors to amino acids.

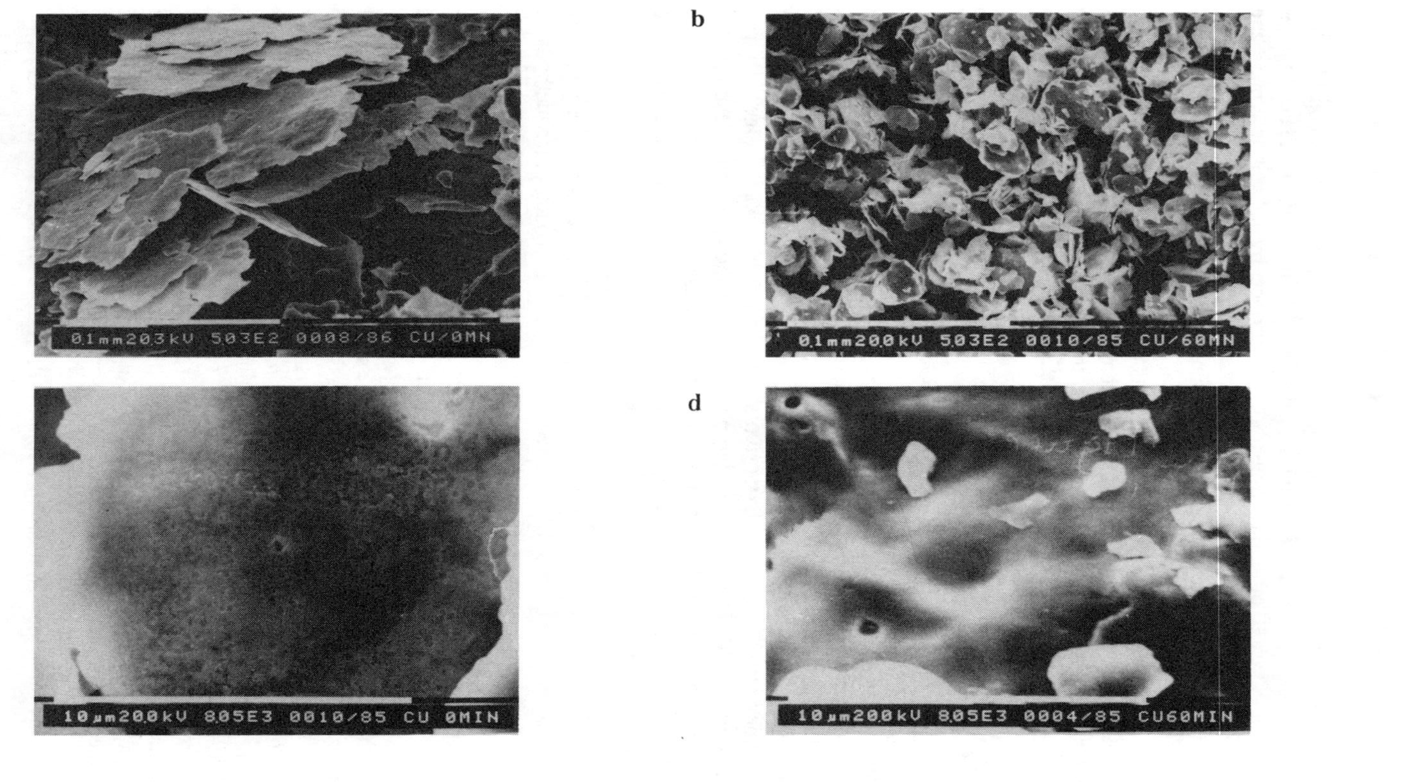

Figure 6. Copper powder (a) unsonicated, (b) after sonication for 1 h, (c) 6(a) at higher magnification, (d) 6(b) at higher magnification.[26]

REACTIONS WITH CONTINUOUS ULTRASOUND

There are a large number of reports of the sonochemical enhancement of reactions involving heterogeneous catalysts/reagents where ultrasound has been used continuously.[32] It is not evident, however, that the ultrasonic conditions have been optimised, since in some cases presonication of the catalyst/reagent is just as effective as with continuous sonication.[33] In several cases prolonged periods of sonication may lead to agglomeration and loss of active surface.[34] Continuous ultrasound is of benefit in those situations where the reaction products form an insoluble coating on the solid catalyst/reagent,[35] and in liquid-liquid systems continuous ultrasound is often necessary to produce emulsions, thus increasing the interfacial surface area.[31]

There have been several reports of the sonochemical enhancement of reactions of aromatic aldehydes such as the Cannizzaro, Michael Addition, Claisen-Schmidt and Wittig-Horner reactions catalysed by a solid barium hydroxide catalyst C-200 which is prepared by calcination of commercial $Ba(OH)_2 8H_2O$.[36] In a mechanistic study of the Wittig-Horner reaction the active catalytic sites were shown by selective site poisoning to be unhindered strong base centres. Infrared studies indicate that in the thermal Wittig-Horner reaction the aldehyde substrate is adsorbed, whereas under sonochemical conditions it is not. It is suggested that the acoustic agitation would be sufficient to overcome the weak π^2-$5d^0$Ba(II) bond formed during the thermal adsorption. The thermal Wittig-Horner reaction takes place between adsorbed ylide and adsorbed aldehyde giving rise to a more rigid transition state than in the sonochemical reaction. A mechanism involving SET between the strong base site on the catalyst and the adsorbed phosphonate is proposed. Small amounts of water have a dramatic effect on yields in both the thermal and sonochemical reactions; this is considered to be related to the stabilisation of the crystalline lattice. By contrast the sonochemical Canizzaro reaction is found to occur at reducing sites on the barium hydroxide C-200 catalyst.[37]

CONCLUSIONS

The application of ultrasound to heterogeneous reactions involving solid and liquid phases has proved to be widely beneficial. Ultrasound is particularly useful in catalyst/reagent preparation through the ability to influence: nucleation and crystallization, intercalation and impregnation of reagents on to supports. Ultrasound is valuable in the activation of solids through surface cleaning and surface modification; these factors are of potential benefit in catalyst regeneration. The use of continuous ultrasound is of benefit in those processes which are mass transport controlled although in some cases continuous ultrasound may lead to agglomeration and loss of activity.

REFERENCES

1. K.S.Suslick, R.E.Cline Jnr., D.A.Hammerton, J.Am.Chem.Soc. 1986, 108, 5641.

2. B.P.Barker and S.J.Putterman, Nature, 1991, 352, 318.
3. W.Lauterborn and H.Bolle, J.Fluid Mech., 1975, 72, 391.
4. W.T.Richards and A.L.Loomis, J.Am.Chem.Soc., 1927, 49, 3086.
5. O.V.Abramov, 'Advances in Sonochemistry', Ed. T.J.Mason, J.A.I. Press, 1991, 2, 135.
6. O.V.Abramov and V.I.Teumin, 'Physical Principles of Ultrasonic Technology', Ed. L.D.Rozenberg, Plenum Press, New York, 1973, Vol.2, p 145.
7. J.Lindley, Ultrasonics, 1992, in press.
8. Y.Ueda, H.Sekiguchi and H.Shirata, Jap. Patent, 1984, 59-82269.
9. V.M.Zhiznevskii, B.Deben, V.A.Kozharskii and E.N.Mokryi, Prumen.Fiz.Fiz -Khim., Metodov Tecknol.Protsessakh, Ed. N.N.Khavskii, Metallurgia, Moscow, 1990, p 119.
10. K.Ivanov, T.Popov and S.Slavov, Izv.Khim., 1987, 20, 201.
11. T.Popov, D.Klissurski, K.Ivanov and I.Pesheva, Stud.Sur.Sci.Catal., 1987, 31, 191.
12. A.N.Maltsev, Russ.J.Phys.Chem., 1976, 50, 995.
13. K.S.Suslick, S-B.Choe, A.A.Chichowlas and M.W.Grinstaff, Nature, 1991, 353, 414.
14. K.S.Suslick, D.J.Casadonte, M.L.H.Green and M.E.Thompson, J.Chem.Soc.Chem.Commun., 1987, 901.
15. P.Lazlo (Ed.), 'Preparative Chemistry using Supported Reagents', Academic Press, San Diego, 1987.
16. T.Ando and T.Kimura, 'Advances in Sonochemistry', Ed. T.J.Mason, J.A.I. Press, London, 1992, 2, 211.
17. W.Marti and K.Waldmeir, Swiss Patent, 1964, 376886.
18. R.Ranganathan, I.Mathur, N.N.Backshi and J.F.Mathews, Ind.Eng.Prod.Res.Develop., 1973, 12, 155.
19. V.I.Shekhobalova and L.V.Voronova, Vestn.Mosk.Univ.Ser.Khim., 1986, 27, 327.
20. J.K.Kochi, J.Am.Chem.Soc., 1955, 77, 5274.
21. D.C.Nonhebel, J.Chem.Soc., 1963, 1216.
22. I.Tanimoto, K.Kushioka, T Kitigawa and K.Maruyama, Bull.Chem.Soc.Japan, 1976, 52, 3586.
23. M.Kodomari, H.Satoh and S.Yoshiotomi, J.Org.Chem., 1988, 53, 2093.
24. B.Flockhart, I.M.Sessay and R.C.Pink, J.Chem.Soc.Chem.Commun., 1980, 439.
25. K.S.Suslick, D.J.Casadonte and S.J.Doktycz, Solid State Ionics, 1988, 444.
26. J.Lindley, J.P.Lorimer and T.J.Mason, Ultrasonics, 1987, 25, 45.
27. K.S.Suslick and D.J.Casadonte, J.Am.Chem.Soc., 1987, 109, 3459.
28. E.A.Cioffi, W.S.Willis and L.S.Suib, Langmuir, 1988, 4, 692.
29. E.A.Cioffi, W.S.Willis and L.S.Suib, Langmuir, 1990, 6, 404.
30. M.Honda, Jap. Patent, 1991, 03000135, Chem.Abstr.114:151432.
    P.G.Rodewald, US Patent, 1990, 4914256, Chem.Abstr.113:26714.
31. R.S.Davidson,'Chemistry with Ultrasound', Ed. T.J.Mason, SCI-Elsevier, London, 1990, p 91-98.
32. J.Lindley and T.J.Mason, Chem.Soc.Rev., 1987, 239.
33. T.Kimura, M.Fujita and T.Ando, Chem.Lett., 1988, 1315.
34. K.S.Suslick and S.J.Doktycz, J.Am.Chem.Soc., 1989, 111, 2342.
35. B.Pugin and A.T.Turner, 'Advances in Sonochemistry', Ed. T.J. Mason,

JAI Press, London, 1990, 1, 81.
36. J.V.Sinistera, A.Fuentes and J.M.Marinas, J.Org.Chem., 1987, 52, 3875.
37. A.Fuentes, J.M.Marinas and J.V.Sinistera, Tetrahedron Lett., 1987, 28, 2497.

NMR Spectroscopy with Ultrasound

J. Homer, S.U. Patel, and M.J. Howard
DEPARTMENT OF CHEMICAL ENGINEERING AND APPLIED CHEMISTRY, ASTON UNIVERSITY, ASTON TRIANGLE, BIRMINGHAM B4 7ET, UK

1 INTRODUCTION

Probably the major thrust in the scientific use of ultrasound lies in the area that is now popularly referred to as sonochemistry. This paper describes a novel area of endeavour in which samples are irradiated with ultrasound while being investigated by nmr spectroscopy. It should be noted that the work largely involves techniques that are quite different from the well known technique of acoustic nuclear magnetic resonance (ANMR). The initiative embraces two principal areas. The first is concerned with changes to the normal nmr spectra of liquids that are induced by their irradiation by ultrasound. The second deals with the motionally induced effects of ultrasound on the nmr spectra of solids. Some of the experiments must be deemed preliminary, may be without full theoretical explanation, and await further authentication, but are of sufficient interest and novelty to warrant mention.

2 THE EFFECTS OF ULTRASOUND ON THE HIGH RESOLUTION SPECTRA OF LIQUIDS

Spin-Lattice Relaxation Times, T_1

The application of ultrasound to liquids may, in principle, modify normal molecular translational and rotational correlation times. If this happens T_1 for components of the liquids may be changed from their normal values. Such changes have been observed reproducibly[1] but only for liquid mixtures, and at high ultrasound frequencies (MHz). Reductions in T_1 of up to 60% may be detected readily.

Table 1 presents ^{13}C T_1 data for an air saturated mixture of 1,3,5-trimethylbenzene, cyclohexane and chloroform-d. The T_1 values were measured using the DESPOT[2] sequence on a JEOL FX 90Q pulse FT multinuclear spectrometer operating at 22.5

<u>Table 1</u> ^{13}C T_1 values for an air saturated mixture of 1,3,5-trimethylbenzene, cyclohexane and chloroform-d (1:1:2 molar ratio) subjected to ultrasound at 1.115 MHz and various powers.

Ultrasound power/W cm^{-2}	T_1/s[a]			
	=C<	=C< H	$-CH_3$	C_6H_{12}
0	7.9	4.2	4.6	8.8
2	7.5	3.9	4.4	7.8
4	6.9	3.6	4.2	6.5
8	6.4<19%>	3.4<19%>	3.5<24%>	6.1<31%>
19	7.6	3.6	3.8	6.5
38	7.8	3.6	3.9	6.8

[a] Average values from two experiments with an average spread of each pair of measured values of ±0.23s.

MHz and 30 °C. It is readily seen that as the power of the ultrasound (1.115 MHz) increases, the values of T_1 decrease to a minimum and then increase. The initial decrease in T_1 is opposite to that found when the sample temperature is raised electrically[1]. It is noticeable that the percentage decrease in T_1 at the minimum appears to depend on the structural environment of the nucleus studied.

It is possible that the technique may offer a new route to the elucidation of molecular structure and molecular dynamics.

<u>The Possibility of Acoustic Magnetic Resonance in Liquids</u>

It is known that sound at a nuclear Larmor precessional frequency can induce nmr transitions in solids. There has been considerable controversy over whether ANMR may be possible in liquids. There are indications that it may indeed be possible[3-6]. Experiments have, therefore, been conducted on ^{14}N-containing compounds for which ultrasound frequencies could be varied through the Larmor frequency of 6.42 MHz. Figure 1 presents the ^{14}N spectra of acetonitrile[7], in equimolar mixture with chloroform-d, when subject to ultrasound at various frequencies and a power of approximately 2.5 W cm^{-2}. It can be seen that with ultrasound at the Larmor frequency, the ^{14}N signal cannot be seen unlike at the other frequencies. It appears possible that the observation at 6.42 MHz could be due to an effect equivalent to ANMR, but electronic saturation of the nmr spectrometer detection system remains a remote possibility.

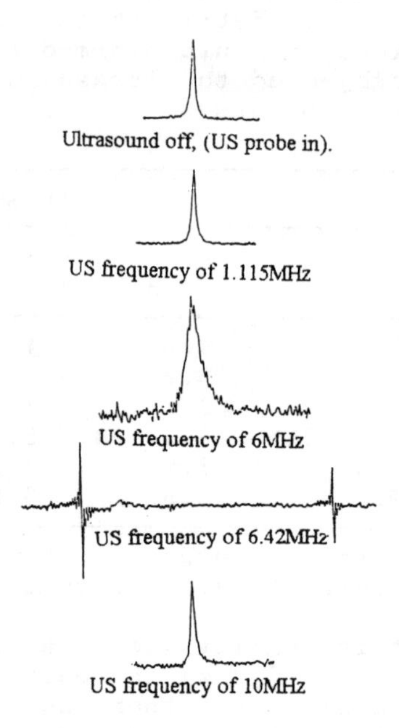

Figure 1 Effect of ultrasound at selected frequencies and 2.5 W cm^{-2} on the ^{14}N resonance of acetonitrile.

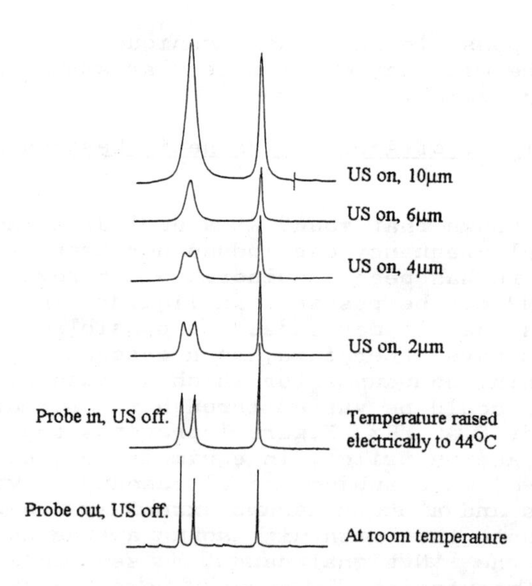

Figure 2 Effect of ultrasonic power at 20 kHz on the ^{1}H nmr spectrum of N,N-dimethylacetamide.

Ultrasonically Induced Conformational Changes

It is well known that at ambient temperatures there is restricted rotation in N,N-dimethylacetamide that enables the observation of three chemically shifted methyl proton resonances. As the temperature is increased the rotational barrier is overcome so that ultimately the two methyl groups attached to the nitrogen atom become chemically equivalent and a single proton resonance is observed.

Figure 2 presents the dependence of the pmr spectrum of N,N-dimethylacetamide on the power of ultrasound applied at 20 kHz[7]. Although the presence of the ultrasonic apparatus results in rather poor spectral resolution it does appear that ultrasound may be used to overcome the intramolecular rotation energy barrier. It is interesting to note that the apparently coalesced spectrum obtained with the highest ultrasound power corresponds to that which would be obtained by thermally heating the sample to well in excess of 100 °C. As will be discussed later, in experiments of this kind it has not been possible to detect sonically induced temperature rises beyond 60°C.

3 SONICALLY INDUCED NARROWING OF SOLID STATE NMR SPECTRA (SINNMR)

Introduction

Towards the beginning of the last decade the senior author proposed that if solid particles were suspended in a suitable support liquid and irradiated with ultrasound it was possible that sufficient rotational motion of the particles might be induced to reduce dipolar and quadrupolar broadening of solid state nmr spectra. If successful the technique could offer a simple alternative to the now classical MAS NMR method, due to Andrew *et al* [8], and the recently reported elegant double rotation (DOR) technique due to Pines *et al* [9].

The first report of apparently successful sonically induced narrowing of solid state nmr spectra (SINNMR) experiments[10] emphasised both the irreproducibility of the technique and the uncertainty regarding its true origin. Subsequent work has concentrated on two main thrusts. The first embraced the elucidation of experimental conditions concerning several interrelated physical parameters necessary for the optimization and reproducibility of SINNMR. The second involved a search for a definitive experiment that could unequivocally be attributed to the proposed origin of the technique. This section of the paper presents in outline some of the observations for peer judgement.

Experimental

All SINNMR experiments have been performed using a high-resolution JEOL FX 90Q pulse FT multinuclear nmr spectrometer. The ultrasound was introduced into samples contained in cylindrical glass or PTFE sample tubes using a titanium horn coupled to a piezo electric transducer driven by a Kerry Ultrasonics Ltd 20 kHz power generator. The horn is 77 cm long with a 19 mm diameter coupling surface and machined exponentially to provide mechanical amplification at the 5 mm diameter probe tip. The power delivered to the sample is determined by the extent of the probe tip displacement, and this was calibrated to show that a 1 µm displacement delivered approximately 6.5 W cm^{-2} to water.

Wherever comparisons were made between static solid and SINNMR spectra identical spectral acquisition parameters were used. For the SINNMR experiments the samples were not spun.

Experimental Parameters and Conditions

The known physical parameters that govern the success of a SINNMR experiment are (a) particle size, (b) particle shape, (c) support liquid density and viscosity and (d) ultrasound power. In relation to the first of these it was speculated[10] that ultrasonic cavitation could result in particle fragmentation to give ultrafine particles that might be subject to rotational averaging by Brownian action of molecules in the support medium[11,12]. This has proved not to be the case for the materials studied to date, and SINNMR spectra have only been observed for particles with dimensions in the range 2 mm to 100 µm and with decreasing efficiency as the particle size reduces. The role played by particle shape is yet to be elucidated fully. So far as the support medium is concerned it has proved necessary to use low viscosity and high density liquids: suitable support liquids are conveniently obtained from mixtures of chloroform and bromoform. Although SINNMR spectra may be observed at various ultrasound powers, a typical value is 3 µm at 20 kHz.

Problems with Identifying the Origin of SINNMR

Possible phenomena that could yield resonance line narrowing similar to the hoped for SINNMR effect are (a) melting, (b) dissolution, (c) molecular motion resulting from phase changes and (d) chemical reaction to produce soluble product species. The last point is particularly important in view of the potential chemical reactivity of the chloroform and bromoform solvent mixes used. The high local temperatures and pressures that can be produced as a result of cavitation during ultrasonic irradiation of samples containing these mixtures could lead to unwanted chemical reactions, possibly involving halocarbenes and halide radicals.

Three sets of experiments illustrating the difficulty in discounting or accounting for these implicit problems are now discussed. The first is included specifically to illustrate the difficulty in establishing the origin of resonance line narrowing from solids subject to SINNMR conditions.

Trisodium Phosphate. The first SINNMR spectra that were obtained reproducibly were for the ^{23}Na and ^{31}P nuclei in trisodium phosphate ($Na_3PO_4 \cdot 12H_2O$) suspended in chloroform/bromoform mixtures. Typical spectra are shown in Figure 3.

Trisodium phosphate has a complex phase diagram[13]. One feature of this is that at 74°C the octahydrate is formed[14] with the possibility of dissolution in the released water occurring. The evidence suggests that this temperature cannot be reached during the course of SINNMR experiments under the conditions employed. Bench experiments reveal that the mixture equilibrates after prolonged sonication at less than 60 °C. Interestingly, the corresponding

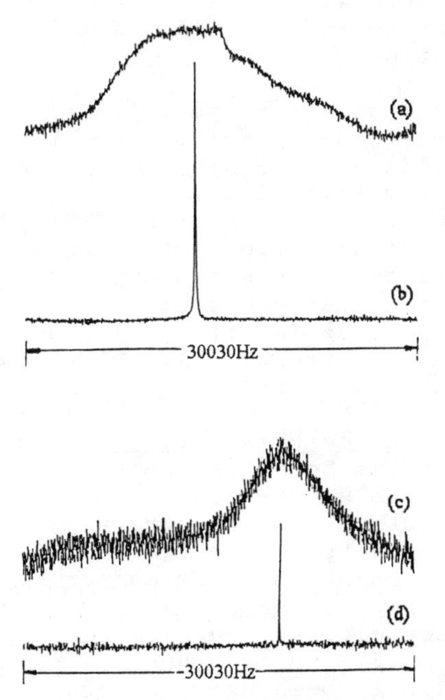

Figure 3 ^{23}Na spectra of trisodium phosphate (a) static solid and (b) suspended solid irradiated with ultrasound at 20 kHz. (c) and (d) are the corresponding ^{31}P spectra.

experiments within the nmr spectrometer show that the equilibrium temperature is significantly less than that achieved on the bench. Superficially, this observation might be attributed to the cooling effect of spinner air in the nmr probe, although this has been questioned by Yesinowski who has made a similar observation[15].

Although effectively insoluble in the organic solvents used, trisodium phosphate is soluble in water, even its own water of crystallization. No evidence of ^{23}Na or ^{31}P containing species has been found in the support liquid, ethanol washings of the solid or from the solid itself, after sonication.

Overall, the evidence suggests that neither melting or dissolution are responsible for line narrowing of the spectra observed. However, it must be acknowledged that there remains the possibility that ultrasound may, in some way, promote unexpected motional behaviour in the otherwise solid matrix that could lead to line narrowing. Nevertheless, there are other observations that indicate that this may not be the case. First, the form of the crystalline material appears to be unchanged after the SINNMR experiment. Second, DESPOT[2] measurements of T_1 for ^{31}P for trisodium phosphate from solution (1M) and SINNMR spectra yielded values of 3 s and 7 s respectively. These are in accord with expectation for this dipolar nucleus in solution and in solid particles undergoing rapid random motion.

Although the observations referred to above may not be considered totally definitive, the implications are that both ^{23}Na and ^{31}P SINNMR spectra may have been observed from trisodium phosphate. If this is indeed the case it appears that SINNMR may enable much improved narrowing of solid state spectra relative to conventional MAS techniques and avoid the production of spinning sidebands that are often typical of MAS spectra. Of additional importance is the fact that SINNMR may enable isotropic shifts for quadrupolar nuclei to be obtained directly.

Polytetrafluoroethylene (PTFE). The ^{19}F nmr spectrum of PTFE has been the subject of extensive investigations. The spectra of static samples reveal significant anisotropy in the screening tensors, so that absorption covers a wide spectral range[16]: absorptions from both amorphous and crystalline regions of appropriate samples may be distinguished. In view of the strong direct dipolar coupling between ^{19}F nuclei in this sample M-REV8 pulse sequences[17] rather than MAS NMR have usually been used to reduce ^{19}F line widths in PTFE samples. PTFE, therefore, offered a severe challenge for the production of definitive SINNMR spectra.

The investigations were conducted on particles cut

from a sheet of highly crystalline PTFE, and suspended in a mixture of chloroform and bromoform. The experimental conditions were initially set by reference to those giving the best SINNMR spectra from trisodium phosphate, and then fine-tuned. The first notable observation during this phase of the experiments was that the commercial bromoform (Aldrich - stabilised by 1-3% ethanol) used is contaminated with a fluorine containing compound that produces a sharp resonance on the lower screening side of the major PTFE ^{19}F absorption, for which spectral parameters were optimized and fixed.

Lines of about 1600 Hz full width at half maximum height (FWHM) were obtained reproducibly by SINNMR. This represents a considerable narrowing relative to the static sample line FWHM of about 3750 Hz (see Figure 4). Although PTFE is unlikely to melt or dissolve under the experimental conditions employed, it must be acknowledged that the material shows phase changes[18] at *ca* 19 °C, 30 °C and 130 °C and that at the higher temperatures chain motion occurs rapidly. For reference, line width narrowing as a function of electrically induced temperature changes was studied. The results are presented graphically in

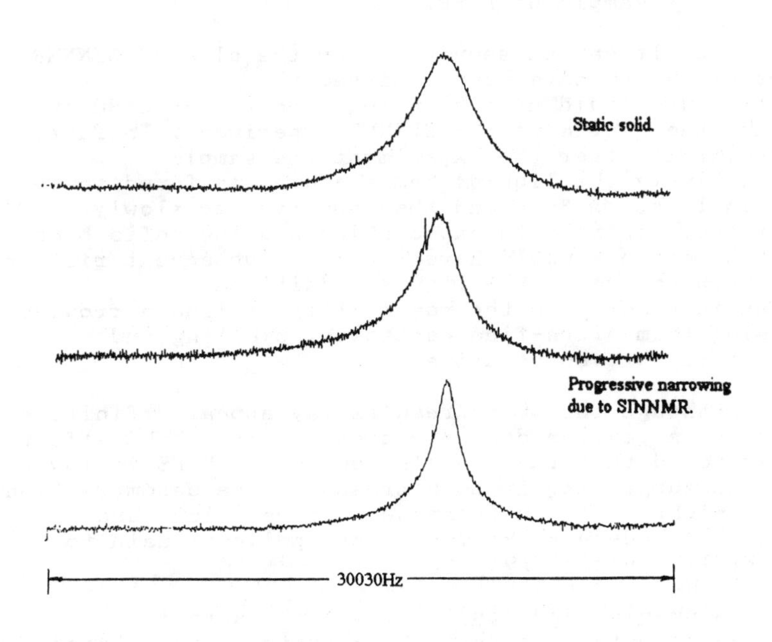

Figure 4 Static solid and SINNMR ^{19}F spectra of PTFE.

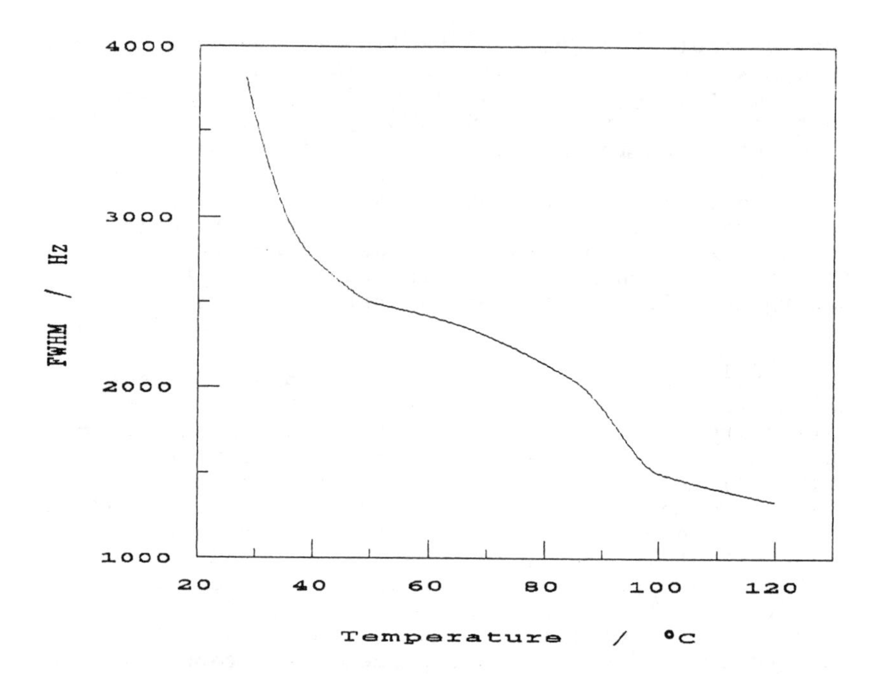

Figure 5 Variation with temperature of the FWHM of the major ^{19}F resonance of a static sample of PTFE.

Figure 5. It can be seen that for the alleged SINNMR line widths to have been achieved the particle temperature would have been required to reach 90 °C during the course of the SINNMR experiment. In fact, immediately after the experiment the sample (effectively the liquid) temperature was found to be initially at *ca* 30°C and then observed to slowly decrease, contrary to expectation had the solid been heated, as this would have caused a subsequent rise in the temperature of the liquid. Additional investigations into the possibility of line narrowing arising from ultra-fine particles, swelling and reactivity proved negative.

Although the above results may appear definitive proof of a genuine demonstration of the SINNMR effect, it is noted that the Young's modulus of PTFE is low and that on subjecting it to ultrasound, the decompression cycle might cause an increase in free volume and facilitate adequate motion of the polymer chain to effect line narrowing.

Aluminium and its Alloys. A characteristic feature of metals is their ability to produce large Knight shifts[19]. The search for a definitive SINNMR experiment led initially to an investigation of the ^{27}Al spectrum of particles of Dural® (typically 95%

Al, 4% Cu and 1% Mg) suspended in chloroform/bromoform mixtures. The unwary should note that when subject to ultrasound aluminium reacts vigorously with the support liquid mixture used. Accordingly, the samples used were coated in a resin matrix.

Andrew[20] has shown that the normally broad ^{27}Al nmr resonance of solid aluminium (FWHM about 9 kHz) may be reduced to about *ca* 700 Hz using the MAS NMR technique. Although the present ^{27}Al studies of Dural revealed a static solid line width well in excess of 9 kHz, the SINNMR experiment revealed a typically Knight-shifted line of about 500 Hz FWHM. There was no evidence of sample melting or any evidence of aluminium containing species in the support liquid after the SINNMR experiment. Moreover, investigation of the solid sample immediately after the SINNMR experiment revealed no Knight shifted, or other, sharp ^{27}Al resonance. Relevant spectra are shown in Figure 6.

Extensive experimental evidence and detailed theory[21] have established that Knight shifts are a consequence of the metallic character of materials. This, in the context of the present results, strongly suggests that the narrow ^{27}Al resonances observed at *ca* 1350 ppm to lower screening than that from aqueous aluminium chloride may be attributed to the SINNMR

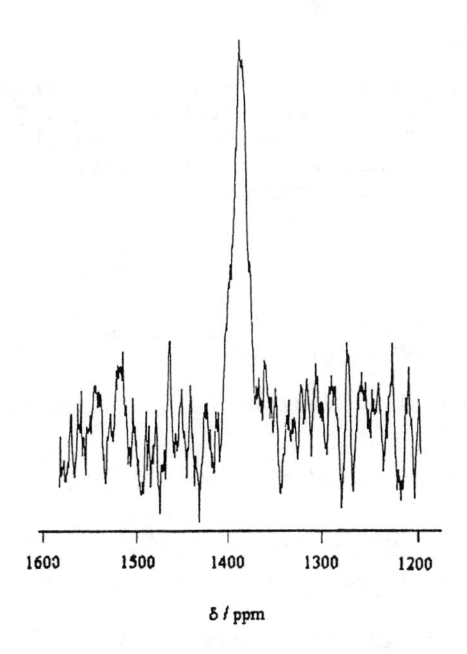

Figure 6 ^{27}Al SINNMR spectrum of metallic aluminium (Dural) relative to aqueous aluminium chloride.

experiment. Soluble aluminium compounds, from whatever source, would produce ^{27}Al resonances at some 1000 ppm, or more, to higher screening than that from metallic aluminium.

4 CONCLUSIONS

It has been demonstrated that ultrasound may be used to change normal molecular dynamics in the liquid phase, and that the induced changes can be monitored by nmr spectroscopy. Additionally, evidence is presented for acoustic nuclear magnetic resonance in liquids.

Although still in its infancy, and considerably more work remains to be undertaken on the subject, it does appear possible to effect the sonically induced narrowing of the NMR spectra of solids. The weight of evidence presently to hand strongly suggests that those spectra purporting to arise from the SINNMR technique may be interpreted to indicate that it does have its origin in the ultrasonically induced random rotational motion of solids suspended in a suitable liquid. Work is now in hand on the investigation of the influence of particle shape and the effect of the frequency of the ultrasound used on the phenomenon, and also on the design and construction of apparatus that will enable its simpler and more widespread use.

Acknowledgments

The authors thank the SERC for financial support, and Mr.P. McKeown, Mr. S. Palfreyman, Mr. G.J. Tilstone, Dr. M.S. Beevers, Dr. E.L. Smith and Professor W.R. McWhinnie for their invaluable contributions to this project.

REFERENCES

1. J. Homer and S.U. Patel, J. Chem. Soc., Faraday Trans., 1990, 86, 215.

2. J.Homer and M.S. Beevers, J. Magn. Reson., 1985, 63, 287.

3. A.R. Kessel, Fiz. Metal. Metalloved,1962, 13, 801

4. A.R. Kessel, Sov. Phys. Acoust., 1971, 16, 425.

5. E.M. Iolin, J. Phys. C., 1973, 6, 3469.

6. E.M. Iolin, Sov. Phys. Dokl., 1974, 18, 527.

7. S.U. Patel, Ph.D. Thesis, University of Aston, 1989.

8. E.R. Andrew, A. Bradbury and R.G. Eades, Nature (London), 1958, 182, 1659.

9. A. Samoson, E. Lipmaa and A. Pines, Mol. Phys., 1988, 65, 1013.

10. J. Homer, P. McKeown, W.R. McWhinnie, S.U. Patel and G.J. Tilstone, J. Chem. Soc., Faraday Trans., 1991, 87, 2253.

11. J. P. Yesinowski, 20th Experimental NMR Conference, Asilomar, California, USA, 1979.

12. K. Kimura and N. Satoh, Chem Lett., 1989, 271.

13. B. Wendrow and K.A. Kobe, Ind. Eng. Chem., 1952, 44, 1439.

14. R.N. Bell, Ind. Eng. Chem., 1949, 41, 2901.

15. J.P. Yesinowski, private communication.

16. A.N. Garoway, D.C. Stalker and P. Mansfield, Polymer, 1975, 16, 161.

17. P. Mansfield, J. Phys. C. Solid State Phys, 1971, 4, 1444: P. Mansfield, M.J. Orchard, K.H.B. Richard, Phys. Rev., 1973, B7, 90: W-K. Rhim, D.D. Elleman and R.W. Vaughan, J. Chem. Phys., 1973, 58, 1772 and 1973, 59, 3740.

18. R.H.H. Pierce *et al*, Meeting of the American Chemical Society, New York, September, 1957: A.J. Vega and A.D. English, Macromol., 1980, 13, 1635.

19. W.D. Knight, Solid State Physics, 1956, 2, 93.

20. E.R. Andrew, W.S. Hinshaw and R.S. Tiffen, Phys. Lett., 46A, 57.

21. A. Abragam, "The Principles of Nuclear Magnetism", Oxford University Press, London, 1961: C.P. Slichter, "Principles of Magnetic Resonance", Harper and Row, 1963.

Ultrasound in Industrial Processes: the Problems of Scale-up

Timothy J. Mason

SCHOOL OF APPLIED CHEMISTRY, COVENTRY UNIVERSITY, PRIORY STREET, COVENTRY CV1 5FB, UK

J. Berlan

ECOLE NATIONALE SUPERIEURE D'INGENIEURS DE GENIE CHIMIQUE, LABORATOIRE DE SYNTHESE ORGANIQUE EN MILIEUX POLYPHASIQUES, CHEMIN DE LA LOGE, 31078 TOULOUSE, FRANCE

1 INTRODUCTION

The problems which are associated with the scale-up of sonochemical processes are mainly related to the estimation of the acoustic energy required to perform a particular transformation and the method by which this amount of energy can be transferred to a large scale process. This requires the definition of a quantity which might be termed sonochemical efficiency relating yield to energy input. In this paper we will attempt to define efficiency, examine those parameters which significantly effect the sonochemical yield and look at some examples of scale-up and the equipment currently available to the process chemist.

For many years power ultrasound has been of interest to chemists as a processing aid in a wide variety of applications (Table 1). More recently the interest of development chemists has been aroused by the prospect of power ultrasonic equipment being made available for large scale chemical plant. Many of the sonochemical transformations identified in Table 1 now have an excellent chance to become industrially significant [1].

Table 1 Uses of Ultrasound in Chemical Technology

SONOPROCESSING	De-Gasification Extraction Crystallisation Emulsification / Dispersion Cleaning and Sterilisation Deaggregation / Particle Size Reduction
CHEMISTRY	Electrochemistry Enzyme Activation Chemical Synthesis Polymer Synthesis and Degradation Heterogeneous and Phase Transfer Catalysis

2 PRELIMINARY SCALE-UP CONSIDERATIONS

When deciding on the type of ultrasonic treatment required for a particular chemical process the first question which must be addressed is whether the ultrasonic enhancement is the result of a mechanical or a truly sonochemical process. If it is mechanical then ultrasonic pre-treatment of a slurry may be all that is required before the reacting system is subjected to a subsequent conventional type reaction. If the effect is truly sonochemical however then sonication must be provided during the reaction itself. Whichever type of ultrasonic treatment is required there are a limited number of ways in which ultrasound can be introduced into the system (Table 2).

Table 2 Methods of Introducing Ultrasound into a Reactor.

1. Immerse reactor in a tank of sonicated liquid.
   (e.g. flask in a cleaning bath).

2. Immerse an ultrasonic source directly into reaction medium.
   (e.g. probe dipped into a reaction vessel).

3. Use a reactor constructed with ultrasonically vibrating walls.
   (e.g. using the tank of a cleaning bath as the reactor)

One of the first steps will be to establish the optimum conditions for sonication in terms of the variables which influence cavitation (Table 3). From the chemical viewpoint the type of reaction involved will also have a distinct bearing on the sort of sonicator system employed. Thus the physical properties of the reaction medium itself will dictate the power of ultrasound necessary. High viscosity media with low vapour pressure will require higher energy to produce cavitation. The presence of entrained or evolved gases will facilitate cavitation as will the presence or generation of solid particles. External parameters are also important when attempting to optimise the reaction conditions. The effect of overpressure on a reaction has been investigated by Cum [2] and the influences of reaction temperature and applied power were reported by Luche [3]. For many years frequency was not thought to be influential in sonochemistry; however there is no doubt that, from recent research, the use of 500kHz irradiation has a significantly greater effect on the generation of hydrogen peroxide from water than 20kHz [4] . In addition to these considerations it is to be expected that the size and the geometry of the chemical reactor employed will have a great bearing on the efficiency of sonochemical activation.

Table 3 Factors which influence cavitation.

viscosity and vapour pressure of the medium
presence of gases or solid particles
overpressure
reaction temperature
sonic power and frequency
size and geometry of chemical reactor

In essence it would be expected that in order to optimise a sonochemical yield there will be an optimum value for almost all the variables which influence cavitation. In addition one might expect that these optima will be very precise so that great care will be needed when scaling up sonochemical processes.

3. THE DESIGN OF SONOCHEMICAL REACTORS

Since a sonochemical reactor should in essence convert electrical energy into chemical reactivity one might expect some energy losses to accrue in driving a sonochemical reaction. Initially there will be electrical and heating losses as electrical input drives the mechanical motion of the transducer through the generator. In turn the transducer motion must be transferred to the liquid medium involving coupling losses. The sound energy then produces cavitation and chemical effects; however the generation of cavitation bubbles also involves heating losses. Finally there will be attenuation of sound energy through the medium by bubbles or suspensions. Also energy will be absorbed (and reflected) from stirrers, baffles, cooling coils or any other devices in the reaction vessel.

4 SONOCHEMICAL YIELD

Somehow an assessment must be made of the efficiency of a sonochemical reactor. This can be defined in terms of sonochemical yield:

Sonochemical Yield = "measured effect" / "sonic power"

The "measured effect" is the amount of product generated in a fixed time (in our studies we have used the amount of I_2 liberated from 4% KI containing 10% CCl_4 in 5 minutes. The "sonic power" is the energy entering reactor and is not the same as electrical power consumed. This can be assessed by calorimetry, the initial temperature rise of system when the ultrasonic irradiation is first switched on (these measurements were carried out using the same proportions of CCl_4 as were used in the iodometric determinations). Ultrasonic power units are quoted either as the energy emitted at the surface of ultrasonic device in Wcm^{-2} or the energy dissipated in the bulk of the sonicated medium (in Wcm^{-3}) [5]. This method of assessment was applied to examples of the three types of of sonicator system described in Table 2.

4.1 *Reactor Immersed in a Cleaning Bath.*

A comparison was made of the efficiency of three differently shaped vessels in a Kerry Pulsatron 55 cleaning bath (35kHz) filled with water containing 5% Decon 90 detergent at 25°C (Fig 1) [6]. Calorimetric measurements revealed that the maximum energy was transmitted into a volume of 55 cm^3 using a 100cm^3 round-bottomed flask whereas for 110cm^3 solution a 250cm^3 conical gave the same result as the 100cm^3 round-bottomed flask (Fig 1a). The results for iodine liberation were however quite different. By far the most efficient result was obtained using a 250cm^3 conical flask containing 55cm^3 reaction mixture (Fig 1b). When these results are combined to obtain a measure of sonochemical yield (Fig 1c) the result is a very clear indication that for this assessment the combination of 55cm^3 reaction volume and a 250cm^3 conical flask is best. This result can be rationalised in terms of the larger base area of the vessel which gives the better transmission of energy and the depth of the liquid in the vessel which is related to the wavelength of sound in the medium.

4.2 *Ultrasonic Probe Immersed in the Reaction.*

In this estimation a Sonics and Materials VC600 sonicator system (20kHz) was used to irradiate a 27.5cm^3 reaction volume [7]. In this situation the results show that for a fixed volume the most efficient result is obtained using the horn with the largest tip diameter (Fig 2). This is explained in terms of the improvement in energy transfer through a larger face of the irradiating source.

Fig 1 Cleaning Bath - Sonochemical efficiency

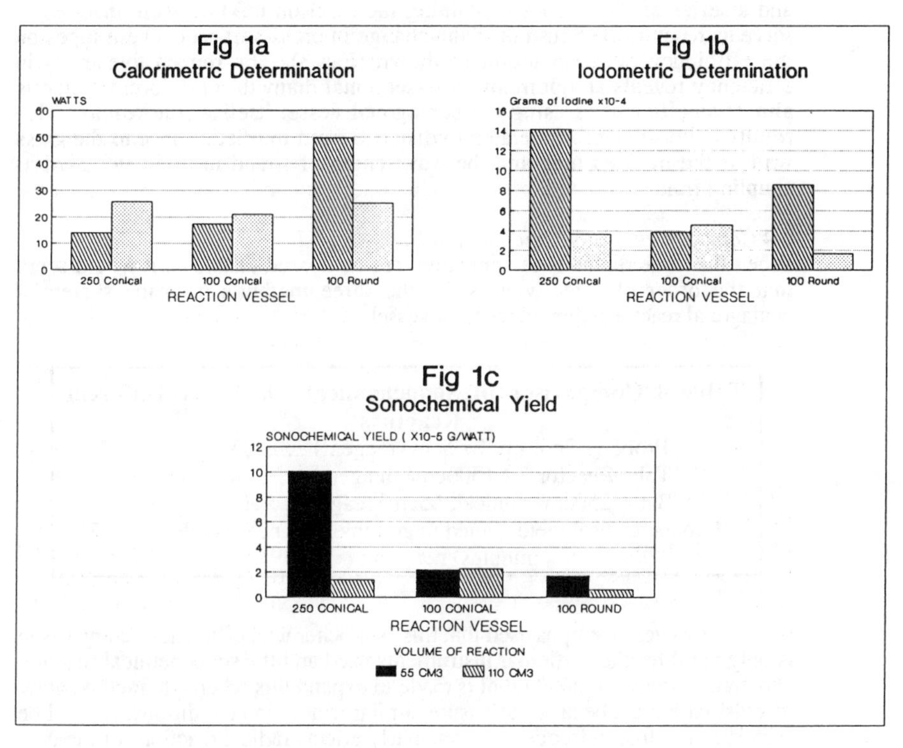

Fig 2 Sonic Probe

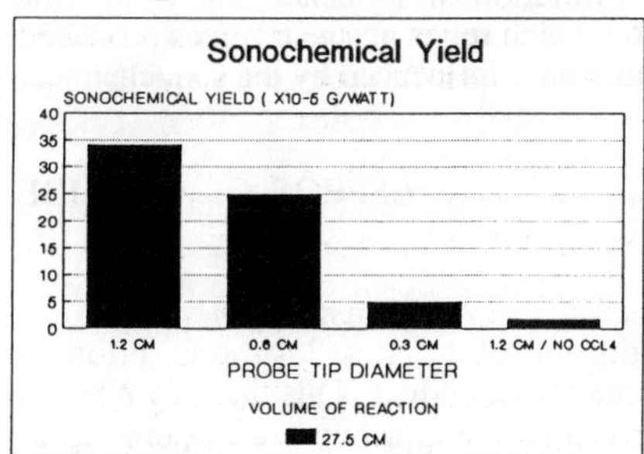

Fig 3 Tube Reactor

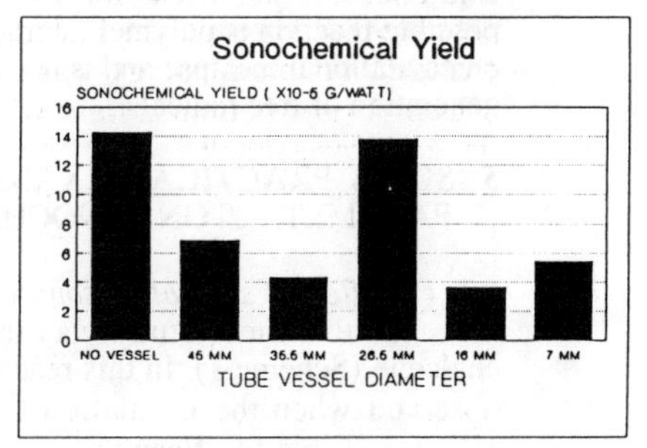

4.3 *Sonication through the walls of a reactor.*
A small Branson pentagonal tube reactor, sealed at one end, with a volume 1500cm^3 operating at 40kHz was first used as a reaction vessel [8]. The pentagonal tube was then filled with water containing 5% detergent (Decon 90) and a series of glass tubes containing the reaction mixture were inserted to serve as reactors. The results of the change in cross-section of these tubes on the efficiency of the reaction is shown (Fig 3). The remarkable trend in efficiency reveals an optimum cross-sectional diameter of 26.5mm which is almost as efficient as using the pentagonal vessel itself as the reactor. This result is almost certainly related to distance from irradiating face to the glass wall of the inserted tube and the wavelength of sound in the water used as coupling fluid.

4.4 *Comparison of sonication methods.*
When the three methods of sonication are compared (Table 4) it would seem that the order of effectiveness for the three methods is probe system > pentagonal reactor > dipped reaction vessel.

Table 4 Comparison of Sonochemical Yields for Different Reactors*

Reactor	Yield
Probe (1.2cm tip, 27.5cm^3 reagent)	35
Tube Reactor (1500cm^3 reagent)	14
Bath (250cm^3 conical, 55cm^3 reagent)	10

* Sonochemical yield quoted in grammes iodine (liberated after 5 minutes) per watt (x10^{-5}).

It must however be emphasised that this "sonochemical efficiency" comparison is only valid for the particular instruments used and the sonochemical reaction chosen as a model. If an attempt is made to expand this scheme to include other reactors care must be taken to ensure similar irradiation conditions. As will be seen later irradiation frequency particularly effects radical reactions of the type used in iodometry and other external parameters may also have a significant influence. For some types of sonochemical reaction it might be advisable to find another model reaction for the estimation of sonochemical yield. One possible reaction is polymer degradation which relies on shear forces produced on cavitational collapse and is not significantly influenced by the sonochemical generation of free radicals.

5 SOME PRACTICAL EXAMPLES OF THE EFFECT OF OUTSIDE PARAMETERS ON SONOCHEMICAL YIELD

5.1 *The effect of sonication source on a Michael Addition reaction.*
The reaction under study was the addition of pentane-2,4-dione anion to chalcone (Scheme 1). In this reaction the initial adduct (1) is the only product observed when the irradiation is performed using a cup-horn operating at 1Wcm^{-3} (Table 5). When this same energy input is applied to the reaction but through a probe system the formation of (1) is accompanied by (2). The latter product results from the cyclisation of (1) and the explanation for this result lies in the difference in power density at the emitting surface.

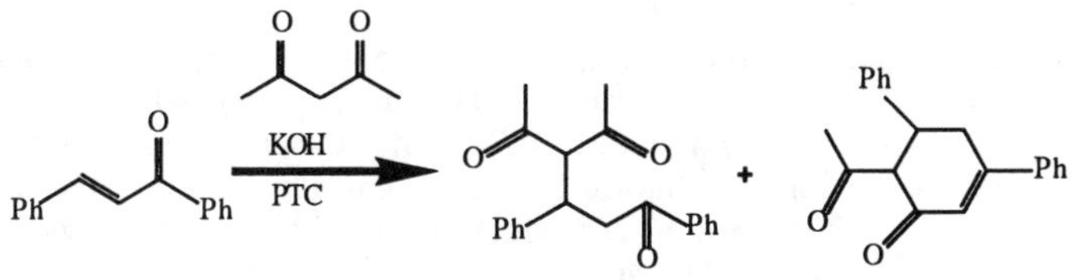

Scheme 1

Table 5 Effect of Sonication Source on a Chemical Reaction.

No Sonication		Cup-Horn[b]		Probe[b]	
1	2	1	2	1	2
52	0	69	0	72	12

(a) after 10 minutes at 25°C,
(b) both systems operating with the same power input of $1 Wcm^{-3}$

5.2 *The effect of sonication source on a "dry" phase transfer reaction.*
The addition of diethylmalonate anion to chalcone was studied under dry solid-liquid phase transfer conditions (Scheme 2). High yields and reaction rates were obtained on a laboratory scale without stirring (Table 6). The scale-up of such a reaction is hindered by the poor heat and mass transfer which are associated with dry conditions. Scale-up would therefore require the presence of a solvent but this results in a reduction in the reaction rate. Fortunately when using a minimal amount of solvent sonication removes the mass transfer problem.

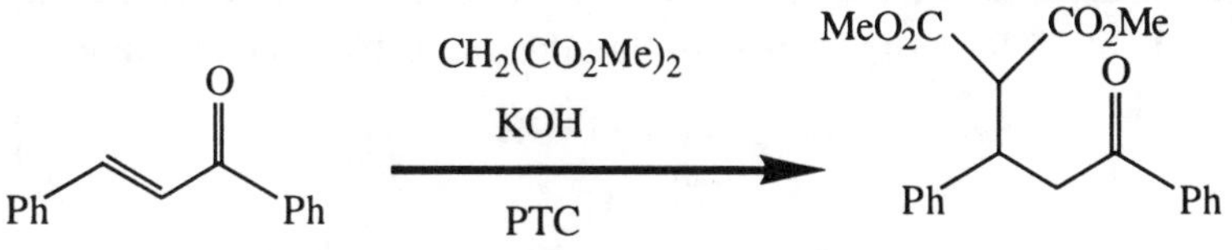

Scheme 2

Table 6 Effect of sonication source on a "dry" phase transfer catalysed reaction.

conditions	reaction time (min)	yield
"dry" non us	5	91
"dry" us	2	97
toluene non us	10	52
toluene us	2	98

(a) using a cup-horn, T = 30°C, power = $1.3 Wcm^{-3}$.

5.3 *Scale-up of a "dry" phase transfer reaction.*
The reaction studied in (b) was scaled-up with the use of a hexagonal ultrasonic bath irradiating a concentric cylindrical vessel of capacity $750cm^3$. The yield obtained in a short time is very high (Table 7). If the hexagonal bath was used in a continuous mode with the cylindrical reactor of capacity $750cm^3$ at a flow rate of $250cm^3 min^{-1}$ then the annual production would be 15 tons based on 6000 hours work. This is probably on a sufficient scale for some fine chemical production in, say, the pharmaceutical industry.

Table 7 Scale-up of a "dry" phase transfer reaction.

Reactor	Yield	Volume
Cup-Horn	97	$50cm^3$
Hexagonal Bath	98	$750cm^3$

Both systems operating at T = 30°C, power = 1.3 Wcm^{-3}, time= 2 min

5.4 *Scale-up of a liquid/liquid phase transfer reaction.*
In this process for the conversion of an organic chloride to a nitrile the reaction medium is water/toluene using an alkali metal cyanide and PTC (Scheme 3). For confidentiality reasons we are not permitted to specify the precise reacting system but two major products are formed - a nitrile and an alcohol and the yields are shown (Table 8). The liquid whistle [8] is an excellent economical choice for the scale-up of processes of this type in that it is very inexpensive and it has low electric energy demand. With a processing volume of $4m^3h^{-1}$ the whistle could produce 300 tons/year of the nitrile.

$$R\text{-}Cl + MCN \xrightarrow{H_2O} R\text{-}CN + R\text{-}OH + \text{byproducts}$$

Scheme 3

Table 8 Scale-up of a liquid/liquid phase transfer reaction.

System	time min	RCN	ROH	Temp
Stirring	90	60	30	40°C (batch)
Whistle	25	57	36	30°C (flow)

(a) Cumulated residence time in whistle chamber (recycling mode)

6 REACTOR TYPES AVAILABLE FOR SONOCHEMISTRY

Thus far in this paper we have referred to several commonly available reactors: cleaning bath, probe system (including cup-horn), hexagonal and pentagonal tubes and liquid whistle. Several other types are also available and can be readily adapted for scale-up (Table 9).

Fig 4 - Submersible Transducer

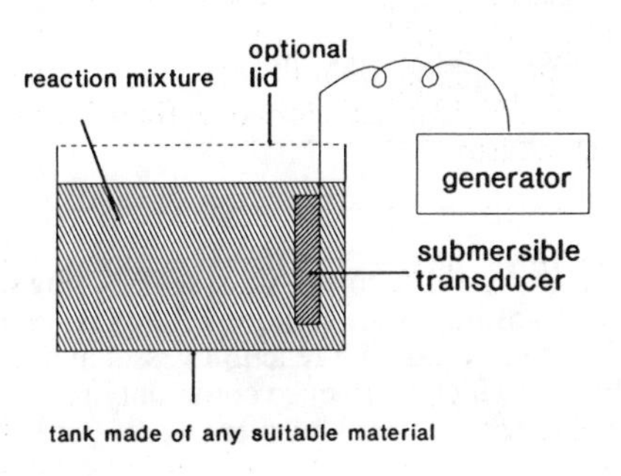

Fig 5 - Flow Cell

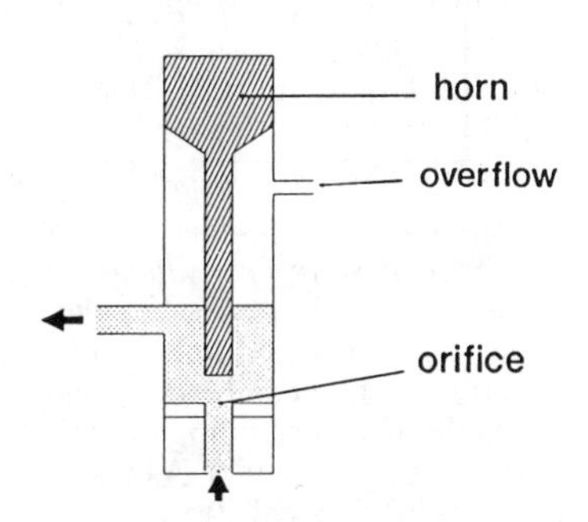

Fig 6 - Cylindrical Reactor

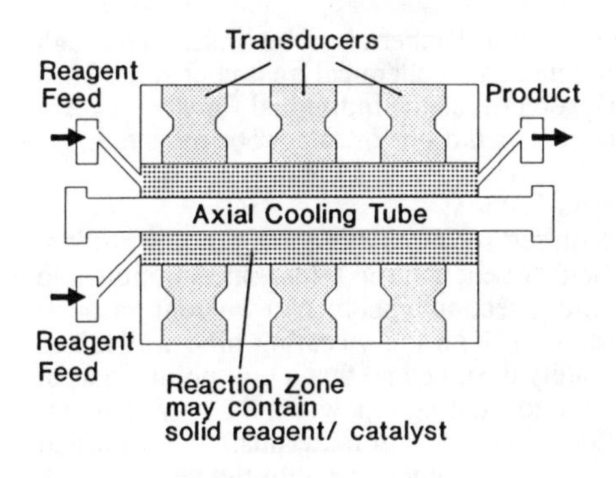

Fig 7 - Nearfield Acoustic Processor

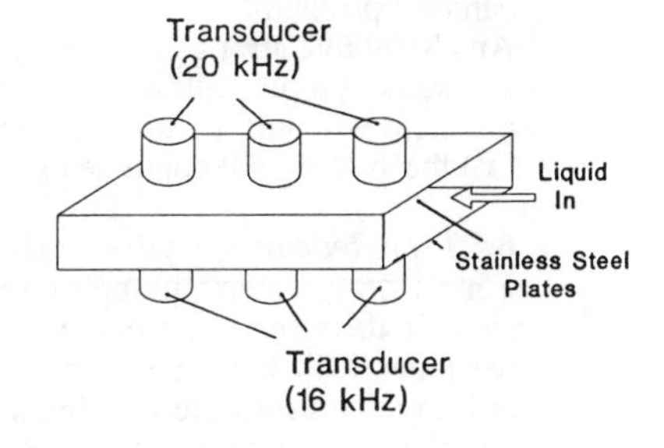

Table 9 Ultrasonic reactors with potential for scale-up.

1.	Submersible Transducer
2.	Probe System with Flow Cell
3.	Tube Reactor (circular cross-section)
4.	Nearfield Acoustic Processer (NAP)

6.1 *Submersible Transducer Assembly.*
Used for many years in the cleaning industry the submersible assembly (Fig 4) consists of a sealed unit within which transducers are bonded to the inside of one face. They can be designed to fit into any existing reaction vessel and are capable of withstanding organic solvents (which were once commonly used in the cleaning industry). As with a cleaning bath system however this reactor will require mechanical stirring.

6.2 *Probe System with Flow Cell.*
Flow systems are generally regarded as the best approach to industrial scale sonochemistry. The general arrangement consists of a flow loop outside a batch reactor which acts as a reservoir within which conventional chemistry can occur. Such an arrangement allows the ultrasonic dose of energy entering the reaction to be controlled by transducer power input and flow rate (residence time). Temperature control is achieved through heat exchange in the circulating reaction mixture. Using a probe system this type of treatment can be achieved on a moderate scale with a flow-cell (Fig 5) although at high powers tip erosion can be a problem.
An alternative arrangement would involve a number of probes inserted through the walls of a pipe allowing a much longer sonochemical treatment zone. Such a system will suffer from the same problems as the individual flow cell except that the system will continue to function even if one or two probe units fail.

6.3 *Tube Reactor of Cylindrical Cross-Section.*
For all flow systems pumping is required for circulation and so they are less suitable for viscous or heavily particulate reaction mixtures. One solution is to use pipes (which can be of various cross-sectional geometry) where ultrasound is introduced into the flowing system through the vibrating pipe walls. The length of such a pipe must be accurately designed so that a null point exists at each end and it can then be retro-fitted to existing pipework. Such systems are capable of handling high flow rates and viscous materials. A cylindrical resonating pipe will provide a focus of ultrasonic energy in the centre of the tube as will a hexagonal pipe. Thus relatively low power at the perimeter inner surface will give high energy in the centre; this reduces erosion problems at the emitting surface. In engineering terms the cylindrical cross-section pipe suffers from the disadvantage that the attachment of transducers to the curved outside surface walls is not so simple as attachment to the planar walls of a hexagonal pipe. One approach is to use a fluid coupling medium [10]. Cylindrical tube reactors with transducers directly bonded to the outer surface have been developed by a number of groups. The Battelle resonating pipe [11] is 6 inches in diameter and operates at 25kHz and has been used for the degassing of oils. The cylindrical reactor developed at the University of Milan [12] has a central core through which a coolant can be passed. In this system a cooling pipe is placed in the focal region of acoustic energy and this acts as both a temperature control and co-axial reflector of the sound waves (Fig 6). The reactor volume is then the annular space between the resonating pipe and the cooling core.

6.4 *The Nearfield Acoustic Processor (NAP)*

Once the concept of a flow loop has been accepted there is no need to restrict the design to tubular reactors. Several alternative designs are available one of which, the Nearfield Acoustic Processor (NAP),was developed some years ago by the Lewis Corporation (Fig 7) [13].

The device consists of two sonicated metal plates between which the reagents flow. The plates are driven at different frequencies (16 and 20kHz) giving an intensity in the liquid between the plates greater than sum of single plate intensities. This processor is driven by magnetostrictive transducers, is very robust and can cope with large throughputs.

7 PROSPECTS FOR SCALE-UP

At this stage in the development of sonochemistry it is clear that a number of different sonochemical reactors are available as direct derivatives of existing systems. These provide high or low sonic power in both batch and flow configurations. New systems are being developed and introduced to the market on a regular basis and it should be only a few years before plants of the future are designed with one or more segments employing power ultrasound as an energy source to enhance chemical reactivity.

REFERENCES

1. T.J.Mason and J.P.Lorimer, "Sonochemistry, Theory, Applications and Uses of Ultrasound in Chemistry", Ellis Horwood, 1989.
2. G.Cum, R.Gallo, A.Spadaro, G.Galli, J.Chem.Soc.Perkin Trans II, (1988) 376.
3. J.C.de Sousa-Barboza, C.Petrier and J.L.Luche, J.Org.Chem. (1988) **53**, 1212.
4. C.Petrier, A.Jeunet, J-L.Luche and G.Reverdy, J.Amer.Chem.Soc. (1992), 114, in press.
5. T.J.Mason, "Practical Sonochemistry, A users guide to applications in chemistry and chemical engineering", Ellis Horwood, 1991.
6. Kerry Ultrasonics, Hunting Gate Wilbury Way, Hitchin, Herts SG4 0TQ, U.K.
7. Sonics and Materials, Kenosia Avenue, Danbury, Connecticut 06810, U.S.A.
8. Branson Ultrasonics, 27 Clayton Road, Hayes, Middlesex UB3 1AN, U.K.
9. Sonic Services, 23 rue Pascal, 75005 Paris, France.
10. Harwell Sonochemistry Development Club, AEA Industrial Technology, Harwell Laboratories, Oxford OX11 1TX, U.K.
11. Battelle, Columbus Laboratories, 505 King Avenue, Columbus, Ohio 43201, U.S.A.
12. Professor V.Ragaini, Università di Milano, Dipartemento di Chimica Fisica ed Electrochimica, via Golgi, 19-20133 Milano, Italy.
13. Lewis Corporation, 102 Willenbrock Road, Oxford, Connecticut 06483, U.S.A.

Sonochemistry: a Chemical Engineer's View

P.D. Martin
ENVIRONMENTAL AND PROCESS ENGINEERING DEPARTMENT, AEA INDUSTRIAL TECHNOLOGY, HARWELL LABORATORY, OXFORDSHIRE OX11 0RA, UK

1 INTRODUCTION

During the past decade sonochemistry has progressed from a laboratory curiosity to becoming an important tool for the synthetic chemist. The degree and variety of interest and activity can be gauged from participation in recent international meetings.[1,2,3] However, if the science and technology of sonochemistry is to be applied in industrial practice the task of scale-up will fall to the chemical engineer. The scale of operation normally envisaged is modest (compared to that encountered by the engineer in the bulk chemicals sector) with typical batch volumes of a few hundred to a few thousand litres, but the challenges are nonetheless significant. Here we examine the issues facing the chemical engineer, compared to those addressed by the chemist, and describe the development of a chemical engineering approach.

Issues Facing Chemists and Chemical Engineers

The concerns of the chemical engineer, though related to those of the chemist, differ significantly, not least in a persistent emphasis on scale and economics; and ultimately it is the chemical engineer who must mediate between chemist and accountant in determining process viability.

In examining a new sonochemical reaction, the *chemist* seeks to answer questions of yield and/or selectivity, as well as potential improvements in operating conditions: it is from these that advantages over conventional reaction techniques derive. He or she will if possible also seek to establish the mechanism whereby these advantages are achieved, and mechanism has an important bearing upon scale-up, as will be seen below.

The chemist will also wish to measure and record the ultrasonic parameters of frequency, total power input and perhaps power distribution. The appropriateness and value of such measurements will depend upon the type of equipment used, typically a laboratory cleaning bath or probe, and the chemist must normally make an off-the-shelf selection of equipment. At present, differences in equipment and methods of recording ultrasonic parameters make it difficult to draw quantitative conclusions from reports of different workers.

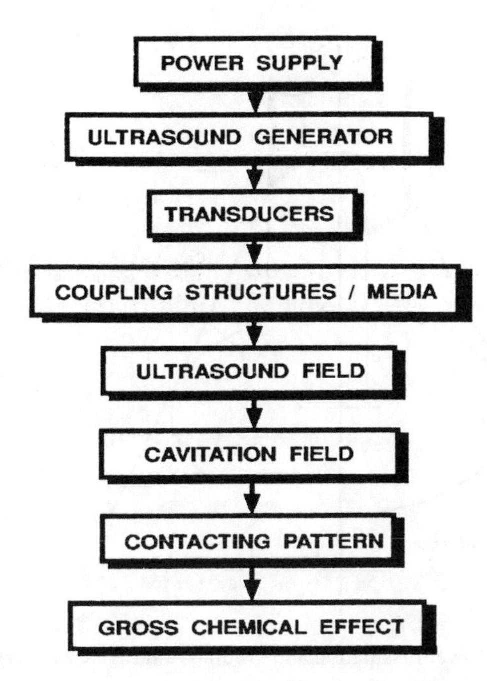

Figure 1 Energy conversion in industrial sonochemistry

This last feature of sonochemical research is of primary importance to the *chemical engineer*. At all scales it will be necessary to the economic case to be able to calculate the volume, intensity distribution and power input to the insonated reagents. The chemical engineer's second question will be whether to attempt to insonate the entire reaction volume or only a part, ie. by bringing the reagents to the insonator: in the latter case it must be possible to describe the overall performance of the reactor in terms of that of its insonated and non-insonated ('silent') parts. The third question concerns equipment, and here it has been found necessary to place the emphasis on design rather than simple selection. The chemical engineer's final concern, as was suggested above, is whether or not the mechanism of sonochemical action permits scale-up with economies of scale.

This quantitative, engineering, approach is illustrated in Figure 1, in which (electrical) power input is related to ultimate sonochemical benefit. At each stage in the transmission of power there is an inefficiency whereby sonochemical benefit is lost to heat or sound.

It is the four questions identified above which are discussed in this paper.

2 ANALYSIS OF THE CAVITATING VOLUME AND APPROACH TO SCALE-UP

In addressing the engineer's first concern, that of correlating ultrasonic parameters, the challenge is to relate the parameter of power distribution or intensity, measured in units of power flux, to the total power effectively (ie. cavitationally) deposited in a given volume. Here we address this question of converting from an area-specific parameter to one which is volume-specific.

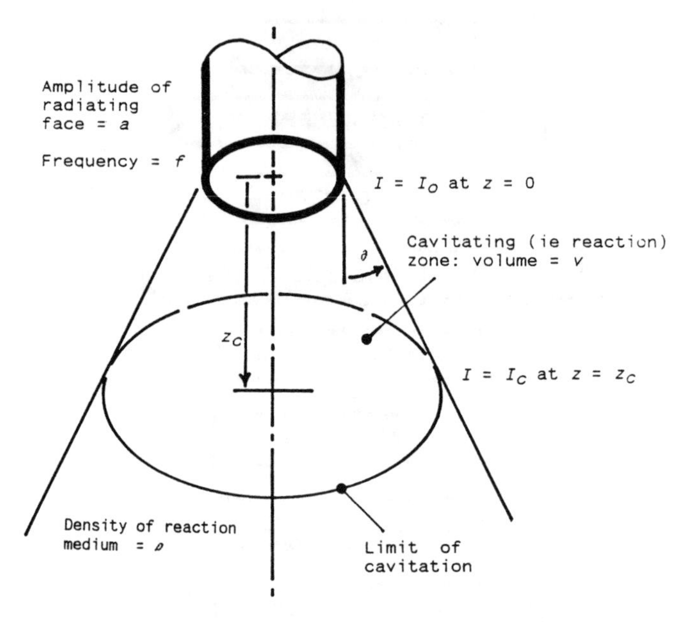

Figure 2 Intensity, power dissipation and volume of the 'active' cavitation zone

Because of the nature of the transducers, the source of the ultrasound field will typically be a circular surface vibrating with frequency f and amplitude a, Figure 2. An intensity $I = I_0$ is delivered by the radiating face, defined by;

$$I = 2\pi^2\rho f^2 a^2 c \tag{1}$$

...and energy is dissipated largely through a 'mainlobe' of the ultrasound field subtending an angle θ from the normal. Intensity falls off with distance from the radiating face by cavitational dissipation and beam spreading, until the intensity falls below the threshold intensity for cavitation: beyond this threshold distance cavitation is absent, and energy is wasted as heat. In practice, at the frequencies most used for sonochemistry, 20-50 kHz, the effective range is a few centimetres, or at best a few tens of centimetres. If Λ is the attenuation constant, it can be shown, using appropriate assumptions, that the power dissipated in the cavitating zone is given by

$$P = \int_0^{z_c} \pi(r_0 + z\tan\theta)^2 \Lambda I_{(z)} dz = \pi I_0 r_0^2\left(1 - e^{-\Lambda z_c}\right) = P_0\left(1 - e^{-\Lambda z_c}\right) \tag{2}$$

The power so dissipated can in turn be related to the measured conversion by means of an empirical constant.

Looking at the above analysis, it can be seen that intensity, power distribution and actively cavitating volume can in principle be correlated, at least for a single source; some uncertainties still remain regarding the interaction of multiple sources.

However, the range of cavitationally effective ultrasound is limited even in homogeneous systems, and will be more so in the many heterogeneous systems of interest to sonochemists. Therefore, in scaling up batch reactors for industrial application some form of two-zone layout is most appropriate. Practically, this will take the form of an internal or external *loop reactor*: this type of reactor will also be suitable for continuous flow schemes if operated in a feed-and-bleed mode.

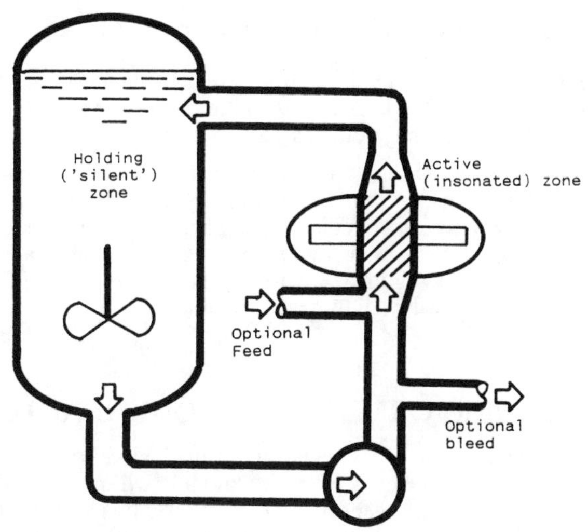

Figure 3 Loop reactor for sonochemistry

The use of a loop reactor, Figure 3, brings a number of advantages. The problem of penetration range is overcome; the residence time of reagents in the actively insonated part can be controlled; and a modular approach can be taken to the design, installation and maintenance of the power ultrasound module, which can also be used alone in the laboratory if so desired.

However, a number of questions need to be addressed in the design of such a loop reactor. Regarding the *whole reactor*, the effects on performance of the volume ratio between the insonated and 'silent' parts needs to be established, as well as the effect of recirculation flowrate: in other words, the effects of the relative and absolute residence times in the insonated zone need to be known.

With regard to the *insonator module*, the main design issue is the operation and performance of multiple transducers. However, serious consideration needs to be given also to maintainability, as well as to the potential acoustic and flammability hazards of industrial operation.

It is useful to address questions relating to the insonator first, before returning to the performance of the whole loop.

3 DESIGN OF POWER ULTRASOUND EQUIPMENT

In the design of power ultrasound equipment for high intensity cavitating duty in a through duct, multiple transducers will normally be required. Some of the technical challenges which must be overcome are as follows:

(i) Transducer mounting

(ii) Coupling to the liquid load

(iii) Controlling interaction between the transducers, within the reaction medium and between each other through the hardware

(iv) Ensuring reasonable transducer life

(v) Providing for selective transducer removal for maintenance.

A number of extant types of ducted insonator, primarily designed for liquid mixing duties, have been found unreliable or unsuitable in practice, precisely because one or more of the above challenges has not been properly resolved.

The Harwell Sonochemical Reactor module[4] consists of a 13 cm diameter cylindrical duct insonated by three transducers disposed radially about the mid-plane; see Figure 4. The transducer assemblies do not contact the reaction medium directly, but via a buffer fluid selected for its very high cavitation threshold. In this way the transducer assemblies are freed from cavitational erosion, and can be maintained independently, if necessary while the rest of the equipment is running.

4 OVERALL PERFORMANCE OF LOOP REACTORS

Having addressed the question of the reactivity of the 'active' (insonated) zone it is necessary to be able to predict how well such a module will perform when linked to a conventional reactor vessel of lower effective reactivity. It is this question more than any other which will determine whether scale-up with economies of scale is likely to be feasible.

In a loop reactor such as that shown in Figure 3 let it be said that the insonated and 'silent' zones with volumes v and V are perfectly well mixed, and that the concentrations of the key reagent leaving them are C_s and C_v respectively: the volume of the pipework is neglected and there is no back- or forward-mixing between the zones.

Figure 4 Harwell Sonochemical Reactor module

In the trivial case of zero-order reaction with rate 'constants' k_{0s} and k_0, the change in concentration C with time t is;

$$C = t(vk_{0s} + Vk_0)/(v + V) \tag{3}$$

For first order reaction, mass balance for flow and reaction gives

$$dC_s/dt = -k_{1s}C_s - Q(C_s - C_v)/v$$

$$dC_v/dt = -k_{1v}C_v - Q(C_v - C_s)/V$$

subject to: $C_v = C_s = C_0$ at t = 0, and $C_v = C_s = 0$ at t = ∞.

Therefore

$$C_v = C_0[((k_1 + \mu)/(\mu - \lambda))\exp\lambda t - ((k_1 + \lambda)/(\mu - \lambda))\exp\mu t] \tag{4}$$

where

$$\lambda, \mu = \left[-\left(k_1 + k_{1s} + \frac{Q}{v} + \frac{Q}{V}\right) \pm \sqrt{\left(\frac{Q}{v} + \frac{Q}{V}\right)^2 + (k_{1s} - k_1)\left(2\left(\frac{Q}{v} - \frac{Q}{V}\right) + (k_{1s} - k_1)\right)} \right] /2$$

In the same way, the performance of a reactor operated continuously, ie in feed-and-bleed mode, can be found. For first order reaction in such a reactor, for which the net throughput flowrate is q and the recycle ratio R, the fractional reduction in concentration of the key reagent will be

$$C_v/C_0 = q^2(R + 1)/((Vk_1 + q(R + 1))(vk_{1s} + q(R + 1)) - q^2R(R + 1)) \tag{5}$$

5 MECHANISM AND SCALE-UP

In a loop reactor in which the reagents are recirculated, it is apparent that the benefits of insonation will be reduced as the relative volume of the 'silent' zone increases with respect to that of the insonated zone. The modular design of the insonator makes scale-up straightforward; but the challenge is to achieve real economies of scale by maximising the effectiveness of the insonated zone. The best results can be obtained in the following circumstances;

(i) The residence time of the limiting reagent in the insonated zone is optimised, by control of flowrate.

(ii) The limiting reagent, such as the solid in an heterogeneous reaction, is retained in and around the insonated zone; thus the limiting reagent is continuously insonated, or at least its effective v/V is much greater than for the other reagents. This strategy has been successful in the case of the Barbier reaction, where the lithium reagent can be retained in a cage within the insonated zone; Figure 5.

(iii) One or more of the reagents remains activated for a period of time after it has passed through the insonation zone, that is to say it retains a 'memory' of insonation.

In predicting the performance of large reactor loops, retention of activity can be modelled by simple formal means, by assuming exponential decay of the rate constant with time, from k_{1s} to k_1 for instance, from the point at which activated material leaves the insonator. Figures 6a and 6b show the effect of incorporating 'memory' into

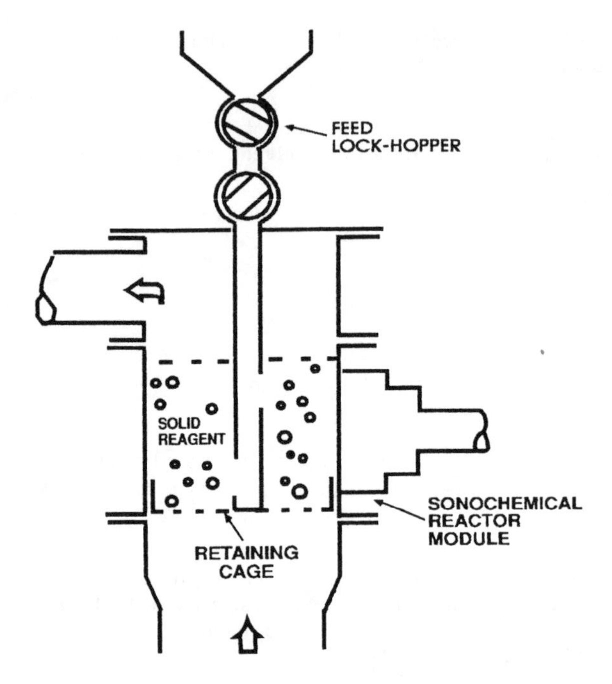

Figure 5 Retention of solid reagent or catalyst in the 'active' zone

reactor models, for batch and continuous flow operation. In each case the volume v of the insonated zone is 4 litres, and examples are given in which the 'silent' volumes V are 0, 0.1, 1.0 m^3, and 1.0 m^3 with 'memory'.

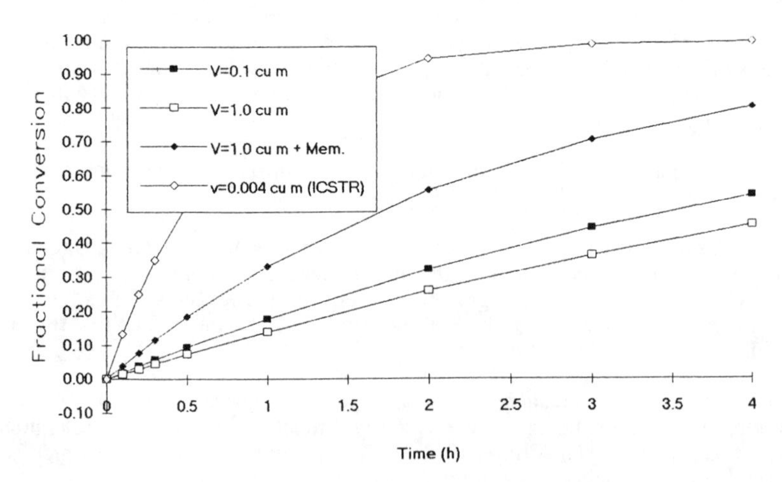

Figure 6a Performance of batch reactors with and without retained activation ('memory')

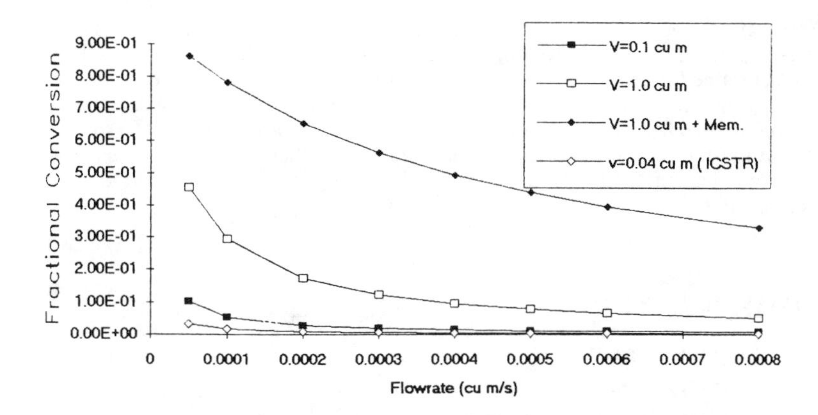

Figure 6b Performance of continuous reactors with and without retained activation ('memory')

Significance of Sonochemical Mechanism to Economies of Scale

It is important to consider which types of sonochemical reaction mechanisms are likely to be most susceptible to retained activation, and therefore to efficient scale-up. Proposed mechanisms can be divided into:

(a) Electrical;

(b) Physical, 'hot spot'; and

(c) Mechanical.

It is reasonable to expect that the effects of the electrical and 'hot spot' mechanisms will be too short-lived to lead to retained activation, unless their effect is either to modify the course of the reaction or to generate active species which are self-propagating to a significant extent. However, mechanical activation, resulting for instance from surface damage or cleaning, could be retained for many seconds or minutes.

6 CONCLUSIONS

The application of chemical engineering principles to sonochemistry has demonstrated a strategem for bringing the technology out of the research chemistry laboratory into industrial practice. The limited penetration range of ultrasound in practical liquid media, as well as considerations of maintenance and modularity, led to the use of loop reactors as the preferred route to scale-up.

Power ultrasound equipment is now available specifically designed for sonochemical duties, and is capable of delivering high intensity cavitation over a volume which is industrially useful in a pilot or small industrial loop reactor.

The efficiency of operations in a scaled-up reactor will be maximised if: the recirculation rate, and hence residence time in the insonated zone, is optimised; if the limiting reagent is retained in the insonated zone; and/or a recirculated reagent exhibits retained activation after exposure to ultrasound. Retention of activity will in turn depend on the mechanism by which ultrasound operates for the case in question; and a comparison of the timescales of activation suggests that heterogeneous systems which rely upon mechanical means of activation will be most readily susceptible to economies of scale.

NOMENCLATURE

a	Amplitude (m)
c	Velocity of sound (ms^{-1})
C	Concentration of limiting reagent (kmol l^{-1})
C_s	Concentration leaving insonated zone (kmol l^{-1})
C_v	Concentration leaving 'silent' zone (kmol l^{-1})
C_0	Initial or feed concentration (kmol l^{-1})
f	Frequency (s^{-1})
I	Ultrasonic intensity (Wm^{-2})
I_c	Threshold intensity for cavitation (Wm^{-2})
I_0	Intensity at radiating face (Wm^{-2})
$k_{n(s)}$	Rate 'constant' for nth. order reaction (sonochemical) ($kmol^{1-n}m^{3(n-1)}s^{-1}$)
P	Power (W)
P_0	Power delivered at radiating face (W)
q	Volume rate of feed and bleed (m^3s^{-1})
r_0	Radius of radiating face (m)
R	Recycle ratio (-)
v	Volume of insonated zone (m^3)
V	Volume of 'silent' zone (m^3)
x	Fractional conversion (-)
z	Perpendicular distance from ultrasound source (m)
z_c	Critical distance to cavitation threshold (m)
θ	Included half-angle of (mainlobe) ultrasound beam (rad)
λ	Parameter defined in Equation 4
Λ	Absorption coefficient (m^{-1})
μ	Parameter defined in Equation 4
ρ	Density (kgm^{-3})

REFERENCES

1. Proceedings *Ultrasonics International 89 Conference*, Butterworths, 1989
2. J.-L. Luche, T.J. Mason and G. Cum (Eds.), *Proceedings 1st Meeting European Society of Sonochemistry*, Autrans, France, 1990
3. V. Ragaini, G. Galli, J.-L. Luche and T.J. Mason (Eds.), *Proceedings 2nd Meeting European Society of Sonochemistry*, Gargnano, Italy, 1991
4. C.L. Desborough, R.B. Pike and L.D. Ward, *UK Patent Application* GB 2 243 092 A; *European Patent* EP 0449 008 A

Sonochemistry: Current Trends and Future Prospects

Timothy J. Mason
SCHOOL OF APPLIED CHEMISTRY, COVENTRY UNIVERSITY, PRIORY STREET, COVENTRY CV1 5FB, UK

1 INTRODUCTION

Sonochemistry has moved on a long way from its beginnings in the 1940's and 50's and through its renaissance in the 1970's to where it stands today as a major new technology available to chemists and chemical technologists. Not only has the subject broadened in scope and gained acceptance in chemical engineering for scale-up, it has also been found particularly beneficial when applied jointly with another technique. Such joint applications have resulted in major advances in these specialisms, two examples being ultrasound with electrochemistry (sonoelectrochemistry) and ultrasound in biotechnology, particularly in the field of enhanced enzymatic reactions. In this overview both of these areas will be explored together with a few other areas which show real potential. The recent interest in sonochemistry has been accompanied by somewhat of a surge in the development of new equipment for the generation ultrasound. Some studies have been taking place outside of the frequency range normally used in sonochemistry (20 to 100 kHz). These offer such promise that it seems likely that we require a new and wider definition of sonochemistry as the uses of sound in chemistry. This would encompass all sound frequencies from infrasound (below 16Hz) through audible sound and all the way through to diagnostic ultrasound (above 1 MHz).

2 CURRENT TRENDS

There are some areas of sonochemical research which seem to offer substantial improvements to existing methodologies

2.1 *Electrochemistry with ultrasound*

The superposition of ultrasonic irradiation on an electrochemical process offers several major advantages over conventional technology (Table 1).

Table 1 Advantages of using ultrasound in electrochemistry

Advantages
Keeps electrodes clean
Degasses the electrode surface
Improves mass transport to electrode
Disturbs diffusion layer and stops ion depletion

These advantages lead generally to more efficient processes [1]. In electroplating the

results are quite astounding with improvements in the metal coatings which include better adhesion, hardness and brightness while the process itself has a lower plating current and a higher deposition rate [2].
In the conventional electrolysis of aqueous systems the degassing can be particularly useful for removing a water soluble gaseous product. Consider, for example, the generation of hydrogen and chlorine from hydrochloric acid (Table 2) [3]. Under the influence of ultrasound this system operates with a 3% increase in current efficiency and an enhanced liberation of chlorine. The obvious potential for such a system lies in the ultrasonically induced stripping of corrosive gas from anodic solutions.

Table 2 Gas evolution(cm^3) during the sonoelectrolysis of 22% HCl

	Conventional	With us
Yield H_2	58.3	59
Yield Cl_2	< 1	59.7

gas yields after 6 mins, carbon rod electrodes, 30kHz u/s, 2 Wcm^{-2}

In the case of classical organic electrochemistry ultrasound offers many advantages. Consider the Kolbe electrolysis of phenylethanoic acid in methanol which, under conventional conditions, does not work without the addition of pyridine. The pyridine prevents the formation of a coating on the electrode which stops the passage of current. Ultrasound improves current efficiency in this process but, more importantly, in the absence of pyridine but in the presence of ultrasound there is no electrode fouling (Table 3) [4].

Table 3 Kolbe sonoelectrolysis of phenylethanoic acid

	13% pyr no us	13% pyr us	no pyr us
$PhCH_2OMe$	21	28	32
$PhCH_2CH_2Ph$	60	51	53

Pt electrodes, 100$mAcm^{-2}$: 7.9V without u/s, 6.6V with u/s

Electrochemiluminescence is a technique which is becoming important in bio-assay studies. Electrochemical oxidation of the tris-(2,2'-bipyridine)-ruthenium dication in aqueous oxalate produces the unstable trication which causes decomposition of oxalate with light emission. This emission occurs close to the electrode surface and its intensity is proportional to the concentration of the ruthenium species. As an example of the use of electrochemiluminescence in an immunoassay the ruthenium species would be attached to an antigen but retain its luminescent activity. Exposure to the antibody to be assayed would produce a large complex which would then be shielded against electrooxidation. The diminution of electrochemiluminescent light intensity is a measure of the biomolecule concentration. The sensitivity of this effect can be significantly improved in the presence of ultrasound which causes a doubling in light intensity for the same current passed, a reduction in the cell potential required and reduced edge effects and patchiness of light emission from the platinum flag electrode [5].

2.2 *Ultrasound in biotechnology*

One of the original uses of power ultrasound was to break down biological cell walls to liberate the contents (indeed many horn systems were first marketed as cell disruptors). Subsequently it has been shown that the selfsame power ultrasound can have a positive effect on enzyme activity although the enzymes can be deactivated if the intensity is too high. Examples are to be found in the biochemistry of both whole and disrupted cells. A simple example of the former is the use of low power ultrasonic activation of a liquid nutrient media to enhance the rate of growth and yield of algal cells. This methodology results in a three-fold increase in the level of protein produced by these cells [6].

At higher powers the release of cell constituents is enough to cause remarkable enhancements in enzyme action as is the case with ultrasonic "stimulation" of a suspension of bakers yeast to provide an inexpensive source of sterol cyclase (Scheme 1, Table 4) [7]. This technique provides an enantioselective enzymatic synthesis of a sterol in gramme quantities. There is no effect of ultrasound on the cell-free cyclase system, a result which demonstrates how cell membrane disruption can occur without damage to the contents.

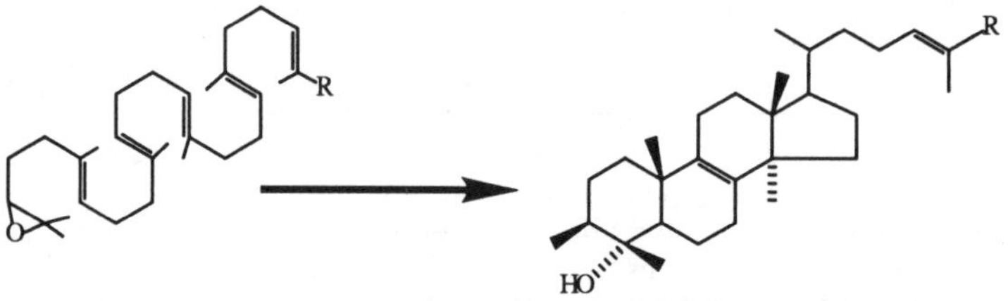

R R HO

Scheme 1

Table 4 Conversion of squalene oxide to sterol with bakers yeast (37°C , 12h)

Enzyme source	conv %	enantiomer conv %
whole yeast	9.5	19
pre-sonicated yeast[a]	41.9	83.9

(a) presonication at O°C using a horn (20kHz) at $40 Wcm^{-2}$ for 2h

Ultrasonically induced emulsification/mixing has been utilised in the two-phase enzymatic synthesis of peptides [8]. Table 5 refers to a biphasic system (38kHz bath) using papain in the dipeptide synthesis shown (eq 2)

BOC-Gly with Phe-N_2H_2Ph ------> BOC-Gly-Phe-N_2H_2Ph ... Eq 2

Table 5 Dipeptide synthesis using water(75%), organic solvent(25%) at 37°C 12h

Organic phase	Stir	Sonicate
diethyl ether	71	89
petroleum ether	12	62

Another fruitful area of research has been that of the sonochemical activation of immobilised enzymes. Using alpha-chymotrypsin (on agarose gel) and casein as substrate a 2 fold increase in activity was observed at 20kHz [9]. The origin of the enhancement was thought to be associated with increased penetration of the casein into the support gel under sonic irradiation. Some inactivation of the enzyme did occur after four repeated uses. With alpha-amylase (on porous polystyrene) an increase in activity with starch was produced on irradiation with 7 MHz ultrasound [10]. At such a high frequency cavitation could not occur and the increased activity in this case is thought to be associated with increased microstreaming of reagents to the surface through the Nernst diffusion layer.

2.3 *Photochemistry with ultrasound*

Historically photochemistry has proved to be a useful synthetic tool normally relying on the electronic excitation of a chromophore to cause reaction. Photochemistry can also be carried out in the presence of a suspension of photoactive material such as TiO_2 where substrate absorption onto a uv activated surface can initiate chemical reactions e.g. the oxidation of sulphides to sulphones and sulphoxides [11]. This technology has been adapted to the destruction of polychlorobiphenyls (PCB's) in wastewater, of considerable interest in environmental protection. Using pentachlorophenol as a model substrate in the presence of 0.2% TiO_2 uv irradiation is relatively efficient in dechlorination [12]. When ultrasound is used in conjunction with the photolysis dechlorination is dramatically improved (Table 6). This improvement is the result of three mechanical effects of sonochemistry namely surface cleaning, particle size reduction and increased mass transport to the powder surface.

Table 6 Photolysis of aqueous pentachlorophenol (2.4×10^{-4} M) containing 0.2% TiO_2

	50 mins	120 mins
uv	40%	no change
uv + u/s	60%	quantitative

2.4 *Regio and stereoselective synthesis with ultrasound*

There is some hope that sonochemical methods might provide improved stereoselectivity of addition reactions, particularly in heterogeneous systems. Consider, for example the C-arylation of (R)- and (S)-2,3-O-isopropylideneglyceraldehyde with phenolates (Scheme 2). For this system syn addition is achieved using Mg^{2+} phenolates (A) and anti addition with Ti^{4+} phenolates (B) [13]. Application of ultrasound (50kHz) to the biphasic solutions results in an increase in both reaction rate and stereoselectivity (Table 7).

(2) Scheme 2 (3)

Table 7 Reaction of phenolates with R-glyceraldehyde acetonide phenolate

	solvent	temp	time	ratio (2):(3)	yield
[ML = $MgBr^+$]	CH_2Cl_2	25	24h	85:15	25% (stir)
[ML = $MgBr^+$]	"	0	5h	96:4	70% (u/s)
[ML = $Ti(OPr^i)_3^+$]	toluene	0	5h	5:95	76% (u/s)

2.5 *Industrial applications*

There are a very large number of industrial patents involving sonochemistry, the one selected here employs ultrasound in two ways. The patent refers to the synthesis of 1-aminoanthraquinone [14]. The initial stage of the process (Scheme 3) involves the oxidation of nitronaphthalene to nitronaphthaquinone using Ce^{4+} in aqueous acidic solution and this is accelerated by ultrasound. The naphthaquinone crystallises out and the aqueous supernatent is recycled. Electrochemical reoxidation of Ce^{3+} to Ce^{4+} is carried out in the recycled liquid but this process produces 3-nitrophthalic acid as by-product. Forced ultrasonic crystallisation of the by-product before recycling prevents its build-up in the main reaction vessel.

NO_2 Ce^{4+} NO_2 O O [-H] NH_2 O O NO_2 COOH COOH Ce^{3+} → Ce^{4+}

If power ultrasound could be efficiently transferred through a gas then it would bring some real benefits to industry such as defoaming and smoke precipitation. Unfortunately there are a number of problems associated with airborne ultrasound which stem from the very high attenuation of sound in air and that a large surface area for the emitter would be required to treat a reasonable volume. Unfortunately this combination is difficult to achieve. If, for example, a normal ultrasonic horn is attached to the centre of a circular tray in an attempt to make a large surface area emitter the result is a cymbal creating a loud noise. Great care is require to develop a disc emitter [15] but the result is well worth the effort. Once such a device is built a number of additional technologies become available. Thus fine powders can "levitate" when subject to an ultrasonic field thus providing a "zero gravity" environment previously accessible only in outer space. It is possible that the airborne ultrasound could also effect gas reactions an interesting area for future research.

3 NEW SONOCHEMICAL EQUIPMENT

There have been a number of developments in the design of equipment for use in sonochemistry some of which might be considered to be spin-offs from the

Fig 1a - Telsonic Tubular Resonator

20 kHz
Transducer
Stainless
Steel Tube
Null Point

Fig 1b - Sodeva Sonitube

Radial
Collar
Stainless
Steel Tube
2 kW
Transducer

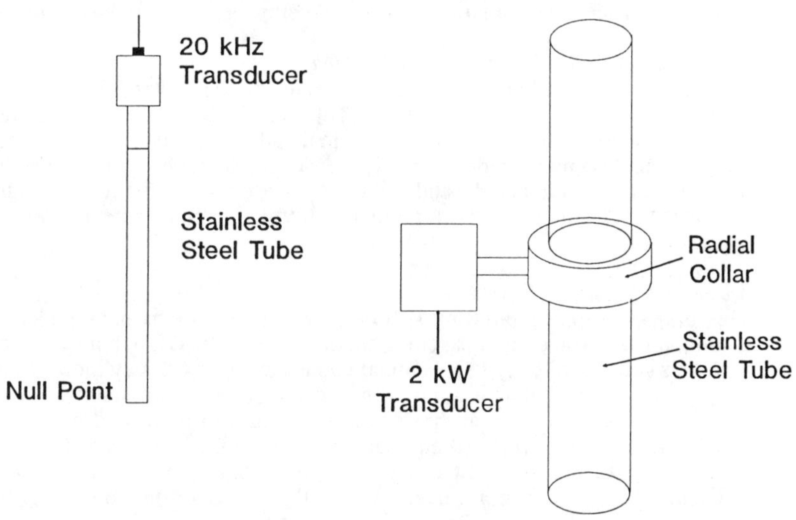

Fig 1c - Martin Walter Push-Pull

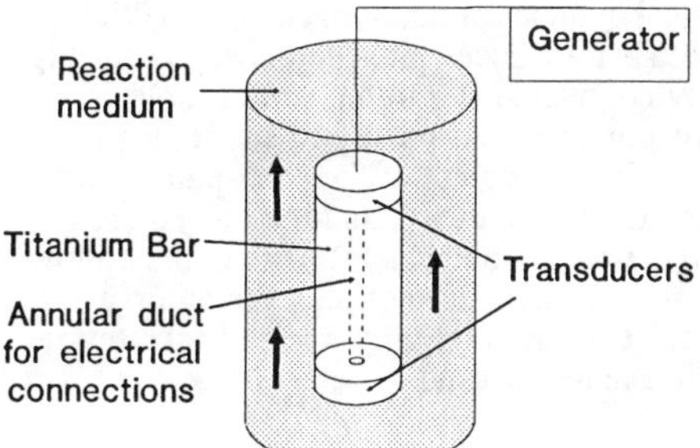

Generator
Reaction
medium
Titanium Bar
Transducers
Annular duct
for electrical
connections

Fig 1d - Undatim Orthoreactor

20 kHz
1 MHz

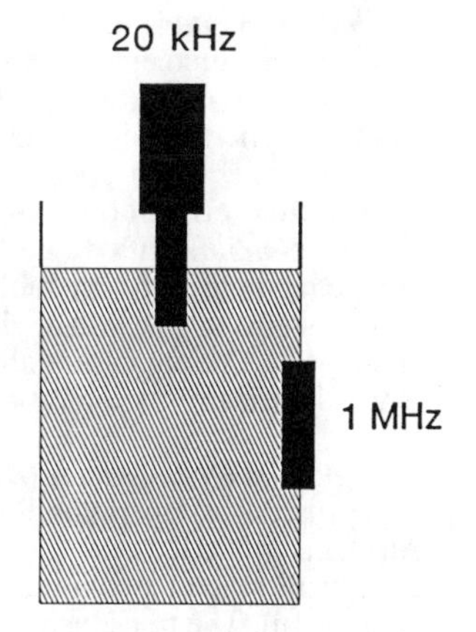

ultrasonic cleaning industry. Four new types are discussed below, the first three of which can basically be classified as potential components for a tube reactor system.

3.1 *Telsonic Tubular Resonator (Switzerland)*
This consists of a hollow, gas filled, tube sealed at one end and driven at the other by a standard piezo transducer (Fig 1a) [16]. This device looks like a conventional probe system but is significantly different in that the sealed end is at a null point and the ultrasound is emitted radially at half wavelength distances along its length. There is the potential to unblock the end and use the system as a flow tube. Designed and marketed for cleaning there is currently no information on its potential for chemical applications.

3.2 *Sodeva Sonitube (France)*
This is also a radially emitting hollow tube but the tube is open ended and the method for ultrasonic input is somewhat different. In this design a transducer horn system is coupled directly to an annular collar which acts as a cylindrical resonator (Fig 1b) [17]. The collar is screw fitted to take one or two stainless steel pipes accurately machined so that the remote end is a null point and may be further coupled to other pipework. At an operating length of 1.2 metres and internal tube diameter of 42 mm the unit can be driven at 2kW per 1.2 metres with 80% efficiency. The maximum resonance power then operates on the flowing liquid at half-wavelength distances along the tube. An advantage for both this system and that of Telsonic is that when the steel tubing becomes eroded it can be easily replaced.

3.3 *Martin Walter Push-Pull (Germany)*
The essential element of the system is a solid cylinder of titanium cut to a multiple of half wavelengths with opposing piezoelectric transducers attached to each end (Fig 1c) [18]. The transducers are electrically connected through a central hole in the cylinder and, when operating the bar responds to the push-pull mode to give a "concertina" effect down its length. The major advantage of the push-pull is that it can be co-axially fitted into the centre of existing pipework and that, if required, it can be made to a considerable length (10 metres is no problem). Unlike the hollow tube systems above this device uses the acoustically more efficient titanium as the radiator and, being essentially a solid bar, it offers a lengthy operational lifetime since significant erosion by cavitation will take a long time.

3.4 *Undatim Orthoreactor (Belgium)*
The principle of operation for this device originates from some studies published by Margulis in the Russian literature [19] and developed by Undatim in Belgium [20]. In a small reactor two ultrasonic fields are introduced at right angles to each other (Figure 1d). The transducer operating in the MHz range generates an ultrasonic field which gives efficient mass transfer. This field also enhances the cavitational effects produced by the higher energy transducer operating in the kHz range. The principle of combining orthoganol ultrasonic irradiation at different frequencies promises significant improvements over irradiation at a single frequency.

4 SONOCHEMISTRY - THE FUTURE

In this last section we will pay a little more attention to cavitation in liquids from which undoubtedly all sonochemistry derives. I hope to show that we should now consider that sonochemistry can be performed over a much wider range of frequencies than that normally quoted.

4.1 *The effect of frequency on sonochemistry*
Historically there have been two underlying misconception in sonochemistry. The first suggests that there is no frequency effect. This misconception arises from the

idea that once the cavitation threshold is crossed the ensuing cavitational collapse will provide the same effects whatever the frequency. The second concerns the frequencies employed in sonochemistry which are generally in the power ultrasound range defined as 20 to 100kHz. This latter despite the fact there is a considerable amount of information on sonochemistry which uses frequencies higher in the ultrasonic range (around 1 MHz) but still involves cavitation [21]. Even if we restrict our considerations to ultrasound in the kHz range there are still some interesting frequency effects. Consider, for example, the well known generation of free radicals in the sonolysis of water. In an important paper on this topic Petrier has compared the effectiveness of 20 and 514 kHz irradiation in the oxidation of aqueous KI to iodine and the generation of hydrogen peroxide in water at the same input power [22]. The rate of production of iodine in oxygen saturated KI (10^{-2}M) was some six times faster and peroxide formation in water 12 times faster at the higher frequency. This result is ascribed to the fate of the OH· radical formed by the breakdown of water on cavitation bubble collapse. The OH· can be destroyed by reactions in the bubble or can migrate into the bulk and produce peroxide. At the higher frequency a shorter bubble lifetime allows more of the OH· to escape from the bubble. We have also some evidence for this frequency effect in work involving OH· detection by fluorescence. When aqueous sodium terephthalate reacts with OH· it forms fluorescent hydroxyterephthalate and its concentration can be estimated spectroscopically [23]. The results in Table 8 show the fluorescence yield after 30 minutes irradiation for 10^{-3}M terephthalate at three different frequencies together with the energy input to the reaction estimated calorimetrically. The sonochemical efficiency for this reaction can be obtained by dividing the fluorescence value (directly proportional to OH· yield) by power input. The results clearly indicate that the efficiency for OH· production increases as the irradiation frequency is increased.

Table 8 Effect of Frequency on OH· production

Frequency(kHz)	20	40	60
Fluorescence	30	40.2	29.3
Power(W)	50	26	11
Efficiency	0.6	1.6	2.7

4.2 *Effect of frequency on sonochemical reactors*

The wavelength of sound in a material is governed by the frequency and the velocity of sound in that material. Consider the consequences of using different sound frequencies in conventional sonochemistry (Table 9).

Table 9 Frequency effects on sonochemical reactor design in conventional ultrasonic range

	20kHz	1MHz
Velocity of sound (Ti rod)	5,080	5080 msec^{-1}
half wavelength for waveguides	12.7	0.25 cm
Velocity of sound (water, 25°C)	1470	1470 msec^{-1}
half wavelength in water	3.7	0.07cm

The consequences of this difference in terms of reactor design are that the transducers employing the higher frequency are much smaller. Unfortunately such devices would provide far less power and penetration than those in the kHz range. For small volume treatment however it is possible to focus the acoustic energy by using a transducer with a concave emitting face and provide a localised high energy region. Chemically the reactions at 1MHz would be expected to generate shorter lived cavitation bubbles giving more radical diffusion into the bulk liquid. There would be more efficient mass transfer to surfaces and less erosion damage.

Why should we restrict ourselves however to this particular frequency range? If sonochemistry is to be connected with cavitation then we should consider any sound frequency which engenders cavitation. Perhaps this precludes frequencies above 2MHz beyond which it is thought that cavitation cannot occur although I believe that this range is of interest in terms of acoustic streaming and mass transport. In cavitation terms however let us consider the audible sound range (Table 10).

Table 10 Half-wavelengths of low frequency sound in metals (metres)

metal	200Hz	20Hz
steel	12	120
titanium	12.7	127

It is quite clear that one of the most important parameters which change as the frequency is reduced is the elongation in resonant wavelength in a metal bar. This can be used to advantage in the case of 200Hz (audible) sound transmitted through steel. In this case we might imagine the simplest concept for scale-up - a giant probe system with a 12 metre bar of steel as the horn. This is precisely the approach of one company [24]. Although the whole device is enormous there is a mechanical benefit from the large (6mm) amplitude vibration at the tip. Such a giant probe system can be used in large scale applications such as the grinding of ores or sawing. When applied to sonochemical processes the immediate possibilities include large scale mixing and the destruction of hazardous waste.

In the case of infrasound frequencies the resonance lengths become unmanageably long - for 20Hz in titanium it is 127 metres. Here we can take another approach and utilise a short length (say 10cm) of metal rod which will behave like a piston when subject to a vibration at this frequency. Such a device has been described by Margulis with the "piston" driving into a reaction mixture contained in a tube. The system is sealed through a plastic collar which transmits the vibrations [25]. The net result of an infrasound energy input is the generation of large deformed bubbles with diameters approximately 1cm across. These have been shown to fragment with the emission of sonoluminescence. Although the sonochemistry for such a system has not yet been widely researched the scale-up potential is clear particularly for processes involving mixing. Undatim[20] intend to introduce such a reactor onto the market in the near future.

5 A NEW DEFINITION OF SONOCHEMISTRY

From the foregoing it is clear that our previous definition of sonochemistry as the effects of ultrasound on chemistry is not correct. It has been shown that cavitation phenomena and hence sound induced chemical changes can be achieved at other frequencies. I would like to propose a new definition of the subject as:

SONOCHEMISTRY - THE EFFECTS OF SOUND ON CHEMISTRY

and within this the ranges involved must be redefined (Table 11).

Table 11 The new sonochemistry ranges

Infrasound	- below 20Hz
Audible	- between 20Hz and 18kHz
Ultrasound	- 18kHz to 100kHz (high power)
	- 100kHz to 2 MHz (low power)
High frequency ultrasound	- above 2 MHz (diagnostic)

6 CONCLUSIONS

Sonochemistry is an expanding field of study which continues to thrive on outstanding laboratory results which have even more significance now that scale-up systems are available. Trends which are now evident include chemical synthesis, sonocatalysis and the use of ultrasound to enhance electrochemistry, biotechnology and photochemistry. The future prospects are likely to encompass a much wider range of applications as the frequency effects are exploited and more studies of sonochemical reaction mechanisms are embarked upon. Equipment design is coming on apace both in the laboratory and in development work for scale-up of sonochemistry and processing. There will be an even bigger effort in sonochemical research as co-operative ventures and technology exchanges are established both within Europe and between Europe and countries of the former Soviet Union. Much of this effort can be linked to the formation, some years ago, of the Royal Society of Chemistry Sonochemistry Group and, more recently, the European Society of Sonochemistry (ESS).

REFERENCES

1. T.J.Mason, J.P.Lorimer and D.J.Walton, Ultrasonics, (1990) **28**, 333.

2. M.R.Walker, Chemistry in Britain (1990) **26**, 251.

3. F.Cataldo, Progega SNC, Rome, private communication.

4. D.J.Walton, A.Chyla, J.P.Lorimer and T.J.Mason, Synth.Comm. (1990) **20**, 1843

5. D.J.Walton, S.S.Phull, D.M.Bates, J.P.Lorimer and T.J.Mason, Ultrasonics, (1992) **30**, 186.

6. Bioprocessing Technology, Technical Insights, U.S.A.(1989).

7. J.Bujons, R.Guajardo and K.S.Kyler, J.Amer.Chem.Soc. (1988) **110**, 604.

8. K.Tadasa, Y.Yamamoto, I.Shimoda and H.Kayahara, J.Fac.Agric., Shinshu Univ. (1990) **26**, 21.

9. Y.Ishimori, I.Karube and S.Suzuki, J.Molec.Catal. (1981) 253.

10. E.Rosenfeld and P.Schmidt, Archives of Acoustics (1984) **9**, 105.

11. R.S.Davidson and J.E.Pratt, Tetrahedron Letts (1983) **24**, 5903.

12. SRI International, 333 Ravenswood Ave., Menlo Park, CA, U.S.A.

13. G.Casiraghi, M.Cornia, G.Casnati, G.G.Fava, M.F.Belicchi and L.Zetta, J.Chem.Soc.,Chem.Commun. (1987) 794.

14. Nippon Shokubai Kagaku Kogyo EP 0 249 969 A2 (1987)

15. Personal communication, J.A.Gallego-Juarez, Instituto de Acustica (CSIC), Serrano 144, Madrid, Spain.

16. Telsonic, Industriestrasse, CH-9552 Bronschhofen, Switzerland.

17. Sodeva, Z.I. de Borly, Cranves-Sales, B.P.7, F-74380 Bonne-sur-Menoge, France.

18. Martin Walter Ultraschalltechnik gmbh, Hardstrasse 13 Postfach 6, Ortsteil Conweiler, D-7541 Straubenhardt 5, Germany.

19. A.F.Dmitrieva and M.A.Margulis, Russ.J.Phys.Chem. (1985) **59**, 1569

20. Undatim Ultrasonics, Parc Scientifique, 10 Rue du Bosquet, B 1348 Louvain la Neuve, Belgium.

21. A.Henglein, Volume 3, Advances in Sonochemistry, JAI press, (1992).

22. C.Petrier, A.Jeunet, J-L.Luche and G.Reverdy, J.Amer.Chem.Soc. (1992) 114, in press.

23. For methodologies used in these estimations see Practical Sonochemistry by T.J.Mason, Ellis Horwood 1991 pp 36-46.

24. Arc Sonics Inc, 5589 Regent Street, Burnaby, B.C., Canada V5B 4R6.

25. M.A.Margulis and L.M.Grundel, Russ. J.Phys.Chem. (1982) **56**, 483.

Subject Index